# MANAGING INNOVATION
## AND CHANGE

# MANAGING INNOVATION AND CHANGE

## People, Technology and Strategy

### JON CLARK

*Professor of Industrial Relations*
*University of Southampton*

SAGE Publications

London · Thousand Oaks · New Delhi

First published 1995

 SAGE Publications Ltd
6 Bonhill Street
London EC2A 4PU

SAGE Publications Inc
2455 Teller Road
Thousand Oaks, California 91320

SAGE Publications India Pvt Ltd
32, M-Block Market
Greater Kailash-I
New Delhi 110 048

**British Library Cataloguing in Publication data**

A catalogue record for this book is available from the British
Library

ISBN 0 8039 8944 X
ISBN 0 8039 8945 8 (pbk)

**Library of Congress catalog card number 95-74574**

Typeset by Type Study, Scarborough
Printed in Great Britain by The Cromwell Press Ltd,
Broughton Gifford, Melksham, Wiltshire

# Contents

# List of figures

# List of tables

# Preface

This is the story of nine years of strategic innovation and change in the UK subsidiary of a multinational company. Between January 1984 and December 1992 the company – Pirelli General, now Pirelli Cables – closed down an old plant, opened a new one 250 km away, and introduced a new human resource strategy which incorporated the latest ideas about the management of people. At its formal opening in 1988 the new plant was described as 'Britain's (and perhaps Europe's) best example yet of twin themes that have been preached for more than 20 years but never realized. The first theme is routine total automation: the second, multiskill employment' (Peter Large, *Guardian*, 4 August 1988).

This book tells the story of how the idea of an automated factory first emerged, and how it was translated into reality. The first part of the story, told in the first two chapters, covers the start of the project in the early 1980s. It deals with the initiation of strategic change up to the point where the company board of directors approved the expenditure for the new project. The second part of the story, told in Chapters 3, 4, 6 and the first part of 8, covers the mid-1980s. It looks at the processes through which senior executives and professional specialists converted the general idea of an automated plant into a detailed blueprint in terms of buildings, layout, machines and machine processes, organization structure, human resource management, and industrial relations. The third part of the story, told in Chapters 5, 7, part of 8, 9 and 10, spans the first six years of operation of the new plant. It traces the implementation, operation and modification of the strategies and policies elaborated in the early and mid-1980s. The fourth part of the story contains a short postscript and a final chapter which draws more general conclusions. These are followed by a personal postscript by Stanley Crooks, one of the leading protagonists in the project and currently Vice-Chairman of Pirelli UK plc.

The story is based on systematic research carried out by the author between 1990 and 1992 (for details, see Technical Appendix). It aims to give the reader a feel of what it is like to work in a commercial organization in the increasingly competitive climate of the 1980s and 1990s. It has three main thematic threads: technical innovation, human resource innovation, and the internationalization of competition and competitive strategies.

This is not a conventional academic book. The main body of the text has no references, footnotes or other academic conventions. It is written largely in a narrative style. The main characters are real people, although to protect anonymity the names used are pseudonyms. It is intended, nevertheless, to be a book from which students – and also general readers – can learn about the management of strategic innovation and change. To this end, each chapter has a main thematic focus. Each chapter also begins with a concise introduction to the main theories, concepts and debates in the research and practitioner literature on the theme in question. Chapter 1 is concerned with the ideas of strategy, structure and organization culture, Chapter 2 with making and justifying large-scale investment decisions, Chapter 3 with locating an experimental plant on a greenfield site, Chapters 4 and 5 with choosing and implementing advanced technologies, Chapters 6 and 7 with designing and implementing innovative human resource strategies, Chapter 8 with designing and operating 'new style' single-union agreements, Chapter 9 with managing quality and continuous improvement (total quality management), and Chapter 10 with leadership and leadership behaviour. At the end of each chapter there is a set of questions for students to discuss, either individually or in groups. The questions are aimed to encourage students to relate the theories and ideas contained in each chapter introduction to the events covered in the main body of the story.

At first sight it might appear that this is simply a story about managers. In order to persuade non-management staff to talk openly and honestly about their experiences, I had to agree to report what they said and did in a way that would not allow them to be identified individually. In contrast, all the managers (and people from outside Pirelli associated with the project) who agreed to be interviewed are identified individually, albeit by pseudonym. In addition, much of the first half of the book – about the early stages of introducing a large-scale innovation – is about the evolution of management strategies. However, as the book progresses, the reader will find that the workforce figures ever more prominently. By the end they emerge, together with a small group of 'leaders', as the central characters in the story.

In a study of developments in human resource management published in 1992, John Storey concluded that the priority for future research was to investigate in detail 'mainstream' companies which have attempted to transform their approach to management and achieved a positive shopfloor response:

> Employee behaviours have, to an extent, been reshaped by the sheer weight of new demands and constraints – for example, where headcounts have been drastically reduced and yet work throughput has remained the same or increased. But the more nebulous aspects of working in a quality way, of offering commitment rather than compliance, and of securing an attitude change to work, to management, to organization and to trade union – all require further investigation. (Storey, 1992: 285)

These questions – and many more – lie at the heart of this book.

# Acknowledgements

The idea for this book originated in 1988 when I was invited by David Yeandle, then personnel manager of Pirelli General, now head of employment affairs at the Engineering Employers' Federation, to help him write two articles for *Personnel Management*, the monthly journal of the British Institute of Personnel Management (now Institute of Personnel and Development). The aim was to outline the innovative human resource policies he and his colleagues had devised for Pirelli's new computer-integrated cable manufacturing plant in Aberdare, South Wales (see Yeandle and Clark, 1989a, 1989b). When I visited the plant for the first time in April 1989, it became clear to me that something quite special was being attempted there, something which was worth investigating further.

At the time, I was looking to re-enter the world of research after two years as head of the Department of Sociology and Social Policy at the University of Southampton. Apart from the fact that the Aberdare site was situated around 250 kilometres from my home, a study of its origins, design and development seemed an ideal project. Here was an exciting attempt to push human resource and technical innovation (the twin themes of my research over the previous nine years) beyond their existing boundaries. For many years the company, whose headquarters was in Southampton, had encouraged close links to its local university and I had known David Yeandle as a colleague and friend for the previous eight years. In the spring of 1990 the company gave me the go-ahead to carry out the project.

During the period of research, I have enjoyed an extraordinary level of cooperation from people inside and outside the company associated with the Aberdare project. Needless to say, my biggest debt of gratitude is to the site managers and workforce at Aberdare for putting up with my visits and seemingly interminable questions between 1990 and 1994. I couldn't have done it without them.

I would also like to thank the Leverhulme Foundation for a personal grant in 1990 to carry out the first stage of the research; the Economic and Social Research Council and the Industrial Relations Research Unit at the University of Warwick for funding a two-year fellowship at IRRU from 1991–3 which allowed me to carry out the second stage of the research, edit a collection of essays on the theme of *Human Resource Management and Technical Change*, and write most of the first draft of this book; and the Department of Sociology and Social Policy and the Faculty of Social Sciences at the University of Southampton for part-funding my stay at Warwick, readily agreeing to free me from teaching and administrative duties for two years, and generously giving me a light teaching load on my return to ensure that I could finish the book before I became Dean of the Faculty in September 1994.

I owe an inestimable debt of gratitude to Stanley Crooks, John Siney, David Yeandle, Keith Sisson and David Winchester, who all supported the project from the very beginning, read the manuscript in draft form, and made numerous helpful comments and suggestions which have been incorporated in the final version. Finally, as ever, thanks to Sue Jones of Sage Publications for her support, constructive criticism and encouragement throughout the project.

*Jon Clark*
*Hoe Gate*

# Dramatis personae

| | |
|---|---|
| **Luciano Balandi** | Head, sector technical department, throughout the design and implementation of the Aberdare project |
| **Paul Bamford** | Seconded to Aberdare project team as accountant (1984–5), carried out investigations into location of new factory (1984) |
| **Gino Brandini** | Aberdare site manager from September 1992; previously member of sector technical department seconded to Aberdare project team (1984–5) |
| **Stephen Cole** | senior Pirelli General manager sent as troubleshooter to Aberdare in early 1989; from June 1989 Pirelli General technical manager; subsequently divisional manager, telecommunication cables division |
| **Martin Dawson** | Aberdare operations manager (1986–90) |
| **Michael Elliott** | Aberdare operations manager (1990–1), site manager (1991–2) |
| **Graham Howells** | Pirelli General employee relations manager (1986–8), then personnel manager (1988–95) |
| **Richard Jones** | Personnel manager, Pirelli General (1960–88) |
| **David Lane** | Aberdare administration manager (1986–9) |
| **Peter McBride** | Aberdare commercial manager (1986–91) |
| **Giuseppe Mancini** | Pirelli General managing director (1985–9) |
| **Alberto Marino** | Pirelli General managing director (1983–4) |
| **Felipe Martinez** | Pirelli General managing director (1989–92) |
| **Martin Parker** | Aberdare maintenance manager (1991–2) |
| **Gordon Peters** | Assisted Barry Roberts in carrying out feasibility study of new general wiring factory (1984); longest serving member of project team to design, set up and develop new factory (1984–94) |
| **Graham Phillips** | Managing director, telecommunication cables division (1991–2), managing director, Pirelli Cables (from 1993) |

| | |
|---|---|
| **David Regan** | Aberdare project manager/divisional manager (1984–91); subsequently commercial director, energy cables division (1991–3) |
| **Barry Roberts** | Conducted feasibility study and produced report on new general wiring factory (1984); from summer 1984 divisional manager of power cables division; subsequently general manager, Pirelli Cables, Australia |
| **Ian Scott** | Divisional manager, general wiring division, based in Southampton (1972–89) |
| **George Sherfield** | Pirelli General accounts manager (1983–91) and, as Pirelli General finance director, member of the Pirelli General board which took the decision to go ahead with the Aberdare project in 1985 |
| **Andrew Slater** | Head of Pirelli General Information Systems Department (1984–91) |
| **Arthur Stokes** | Sector general manager (1982–7) who helped initiate the Aberdare project; subsequently chairman (1987–94) and vice-chairman (from 1994) of Pirelli General, vice-chairman of Pirelli UK plc (from 1989) |
| **David Thomas** | Recruited to Aberdare in 1987 as shift manager, promoted in 1989 to manufacturing manager |
| **Franco Ventrella** | General operations manager, Pirelli General, during the early stages of the Aberdare project |
| **David Wolstenholme** | Seconded to Aberdare project in late 1985, 'systems architect' of the Aberdare factory, died 1990 |
| **Stuart Wood** | Aberdare systems engineer (1987– ), responsible for POMS redesign and implementation (1990–3) |

# MANAGING INNOVATION AND CHANGE

# Initiating Strategic Innovation

## Thematic Focus: Strategy, Structure and Organization Culture

'When I hear the word culture I reach for my revolver.' If Hermann Goering, to whom this quotation is attributed, had worked in a Business School in the 1990s he might have been tempted to substitute strategy for culture. Despite its often over-inflated use, it seems that neither managers nor academics can get by without it. According to a recent survey, there were 37 books in print with the title 'strategic management' (Whittington, 1993: 1), and terms such as 'competitive strategy' (Porter, 1980), 'change strategy' (Kanter, 1983: 381), 'strategic choice' (Kochan et al., 1986) and 'strategic change' (Pettigrew and Whipp, 1991: 25) abound in the management and business literature. There are now few company mission statements which do not refer at least once to strategy and strategic goals. If the word did not exist, we would have to invent it.

Most discussions of the term start with Alfred Chandler's classic book, *Strategy and Structure* (1962). His definition is deceptively simple: 'the determination of the basic long-term goals of an enterprise and the adoption of courses of action and the allocation of resources necessary for carrying out these goals' (1962: 13). For Chandler, strategy is about setting general goals – the overall size and scope of the organization, the mix of products or services it provides, its core values – and deciding on the broad types of action and use of resources needed to achieve them. This *strategy as planning* approach implies that strategies are the intended outcomes of systematic, rational decisions by top managers about clearly defined problems. Strategic change or innovation appears as a linear, sequential process in which strategic analysis and choice (what to do) are followed unproblematically by strategy implementation (how to do it).

Chandler's view of strategy has not been accepted uncritically (see Mintzberg, 1978, 1990; Quinn, 1980, 1988; Pettigrew and Whipp, 1991; Johnson and Scholes, 1993; Whittington, 1993). Mintzberg (1990), for example, has contrasted the idea of strategy as a deliberate, consciously-intended plan with *strategy as an emergent property*, evolving incrementally and piecemeal out of the ideas and actions of people at different levels of the organization. Such strategies may be articulated consciously by top managers, but they are normally rationalizations after the event, a 'pattern in a stream of decisions'. As Pettigrew and Whipp have argued (1991: 31):

> The processes by which strategic changes are made seldom move directly through neat successive stages of analysis, choice and implementation . . . Seldom is there an easily isolated logic to strategic change. Instead, that process may derive its motive force from an amalgam of economic, personal and political imperatives. Their interaction through time requires that those responsible for managing that process make continual assessments, repeated choices and multiple adjustments.

The emergent, adaptive or incremental view of strategy assumes that the internal and external environments of organizations are inherently ambiguous, unstable and unpredictable. However, like the strategy as planning view, it does not assume that managers in organizations can only influence events at the margin, simply adapting pragmatically and opportunistically to continually changing circumstances. The essence of a strategy, and its crucial importance in any process of change or innovation, is that it embodies the '*deliberate and conscious articulation of a direction*' (Kanter, 1983: 294; emphasis added). Successful strategies require both an overall sense of direction *and* a continuous adaptation to change.

In trying to understand strategies and strategy development, it is important to recognize their strong links with organization culture, 'the deeper level of basic assumptions and beliefs that are shared by members of an organization, that operate unconsciously and define in a basic "taken for granted" fashion an organization's view of itself and its environment' (Schein, 1985: 6). Indeed, it is now generally accepted that strategies are both rooted in, and partly explained by, organization culture (see Johnson and Scholes, 1993: 47). The original founders of many of today's large companies – Siemens in Germany, Ford in the USA, Marks & Spencer in the UK, and, as we will see, Pirelli in Italy – played a crucial role in establishing their overall strategy and organization culture (see Schein, 1983; Robbins, 1990). Today, organization culture is one of the most important areas of strategy which can be influenced by top managers and 'visionary' leaders within the organization (see Kanter, 1983: 362; also Chapter 10 for a fuller examination of leadership).

Strategies exist at a number of levels in an organization, of which three are generally distinguished in the research literature: corporate, business, and operational (see Johnson and Scholes, 1993: 10–12). *Corporate* strategy is concerned with the overall size and scope of the organization, its basic goals and objectives, its core values and overall profile, and the general allocation of resources to different operations. *Business* or *competitive* strategy is concerned with the choice of products or services to be developed and offered to particular markets and customers, and the extent to which the choices made are consistent with the overall objectives of the organization. Finally, *operational* strategies are concerned with how different functions within the organization – production or service delivery, finance, marketing, personnel, production, research and development – influence, and are integrated with, corporate and business strategies. The interaction and consistency between the different levels of strategy and structure are a crucial issue for multi-establishment, multi-product and multinational companies.

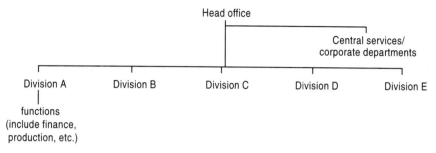

**Figure 1.1** *Multidivisional management structure*

There has been a long-standing debate about the relation between strategy and organization structure. The celebrated phrase of Chandler (1962), that 'structure follows strategy', implies that organizations first plan their strategy, and then design their structure to fit it. In contrast, Mintzberg (1990) has argued that strategies are unlikely to be decided without reference to existing structures. The relationship between strategy and structure is likely to be one of reciprocity rather than one-way determination: 'structure follows strategy . . . as the left foot follows the right' (Mintzberg, 1990, cited in Whittington, 1993: 116). What is beyond doubt is that multinational corporations – the focus of this book – face a number of complex structural problems in developing strategies which are not faced by, for example, small businesses or professional organizations.

Most multinational corporations (MNCs) have some kind of holding or parent company which sits at the apex of the organization structure. Some of these peak organizations are little more than investment banks; others are more centrally involved in developing and overseeing different aspects of corporate and competitive strategy. However, most medium and large companies of today, and almost all multinationals, operate within some form of multidivisional management structure (see Figure 1.1).

In contrast to the more traditional *functional* management structure, in which authority and managerial responsibility are based hierarchically on the main tasks which have to be carried out (production, finance, sales), the *multidivisional* structure or 'M-form' devolves operational responsibility away from headquarters (corporate level) to operational divisions. These are normally subdivided on the basis of different products and product markets and/or different geographical areas. Divisional managers (sometimes called business unit or profit centre managers) are then made directly accountable for the performance of their division. The aim is to allow senior executive managers to concentrate on the development of higher level corporate and business strategies (see on this Whittington, 1993: 112–22; Johnson and Scholes, 1993: 345–56). In multinational companies, the multidivisional structure may operate both at group and national level, making strategy development extremely complex (see on this Johnson and Scholes, 1993: 353–6). For example, different aspects of business strategy may be developed at both multinational group and national company level, one aiming at a global or regional product market, the other at a national one.

What are the main factors which enhance an organization's capacity for strategic innovation? According to Rosabeth Moss Kanter there are five (1983: 289–303):

- departures from tradition
- crises or galvanizing events
- strategic decisions
- individual prime movers
- action vehicles

Almost by definition, innovation and experimentation involve a departure or deviation from traditional or expected ways of doing things. Deviant ideas may emerge in many different ways, but they have a greater chance of getting on the agenda if there is a crisis or problem that is unlikely to be solved by traditional means. Crises or galvanizing events may also represent a challenge to the existing organization culture. They can originate from outside – changes in government policy, competitive environment, product market – or inside – outdated or malfunctioning technology, changes in top management, operational problems – the organization. In most cases innovation will be stimulated by a mixture of these changes.

However, crises and galvanizing events do not of themselves cause change. It is how they are perceived and interpreted by people with organizational power (see Chapter 2), together with the climate of innovation within the organization itself (organizational culture), which ultimately determines whether they act as a trigger for the development of strategies of innovation and change. As Kanter has argued (1983: 306):

> a key to the process of innovation itself is to learn to ask questions rather than assume there are preexisting answers, to trust the process of operating in the realm of faith and hope and embryonic possibility. Repeating the past works fine for routine events in a static environment, but it runs counter to the ability to change. What an innovating organization does is open up action possibilities rather than restrict them and thus trusts to faith as well as formal plans. A well-managed innovating organization clearly has plans – mission, strategies, structure, central thrust, a preference for some activities/products/markets over others – but it also has a willingness to reconceptualize the details and even sometimes the overarching frameworks on the basis of a continual accumulation of new ideas – innovations – produced by its people, both as individuals and as members of participating teams.

It is this mix of planning, the capacity to respond creatively to new developments, and a willingness to take risks, that characterized the early stages of the innovative strategy which led in 1988 to the formal opening of the most experimental project in which Pirelli General, the UK cable affiliate of the Italian multinational Pirelli group, had ever engaged.

# The Royal Opening

In the early afternoon of 22 July 1988, a helicopter landed in a car park just outside Aberdare, a small town in the South Wales valleys. Prince Charles had arrived to officiate at the formal opening of a new manufacturing plant. He was met by the two most influential people in its development, Arthur Stokes, company chairman, and David Regan, the manager of the new unit. After introductions to local dignatories and senior members of the Pirelli group, the Prince posed for photographs before being ushered into the reception area of the new building. After a minute or so talking to the reception staff, he and the invited guests – including members of the national press – entered the factory proper.

It was a massive, windowless, hangar-like structure with machines and machine processes stretching as far as the eye could see. The whole factory was spotless and humming. Continuous lengths of cable were being produced without any sign of direct human involvement. Automatic guided vehicles (AGVs) glided about dalek-like under invisible control, picking up drums of cable from the end of one production line and taking them automatically to the next. On cue, a member of Pirelli staff stepped out in front of an oncoming AGV which was loaded with a large, heavy drum. It had barely touched him before it stopped. No problems with safety there.

For the assembled visitors it was surreal – a factory of the future before their very eyes. It had been described as such in the publicity material inviting them to the opening, but, like so much hype, they had not really believed it. The technology correspondent of the *Guardian* newspaper, Peter Large, a seasoned journalist and author of well-received books on the 'micro-electronic revolution', captured the general mood of the day when he wrote:

> Propaganda has become reality on the outskirts of Aberdare. A . . . factory that at first blush looks utterly boring – it makes household electrical wires – is Britain's (and perhaps Europe's) best example yet of twin themes that have been preached for more than 20 years but never realized. The first theme is routine total automation: the second, multi-skill employment. (*Guardian*, 4 August 1988)

This is the story of how and why the new plant was built and sited at Aberdare, and how managers and staff alike managed the highs and lows of its first five years of operation. Before we begin the story, however, we need to set the context by looking at the structure and strategy of Pirelli General, the company which owned and operated the new plant.

# The Pirelli Group

Pirelli General – now Pirelli Cables – is a wholly-owned subsidiary of the Italian-owned multinational group, Pirelli SpA. In 1988, at the time of the opening of the new Aberdare plant, the group was organized in two industrial holding companies – one (Pirelli SpA) based in Italy, the other

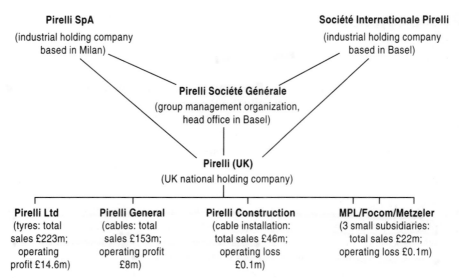

**Figure 1.2** *Ownership structure of the Pirelli group and Pirelli (UK), 1988*

(Société Internationale Pirelli, SA) in Switzerland. Each owned equal shares in all of the group's operating subsidiaries and in Pirelli Société Générale SA, which was responsible for managing the whole group on their behalf. The group was, by any standards, a major force in the global market-place, particularly in its two main product areas, tyres and cables. In 1988, it operated around 140 factories in 16 countries, employing over 70,000 staff with an aggregate turnover of US $6 billion. Sixty per cent of the turnover was still in Europe, but its non-European activities were growing apace.

Each country in which the group operated had a national holding company. Pirelli (UK) was the third largest national company after Italy and the USA. It brought together six product-related subsidiary companies (including Pirelli General) and 11 factories with over 7000 employees and an annual turnover of nearly £500 million. The ownership structure of the Pirelli group and Pirelli (UK) is outlined in Figure 1.2.

When they hear the name Pirelli most people think of tyres (and calendars). However, the company has been involved in cable manufacture since it was founded in 1872 and still retains a major global presence in the cable product market. In 1988, when the Aberdare plant was opened by Prince Charles, it owned 63 cable factories in 14 countries, employing over 20,000 staff with sales of US $2.2 billion, around 36 per cent of its world-wide turnover.

In terms of direct line management, as distinct from ownership, national cable company executives – such as the managing director of Pirelli General – were directly accountable for strategy and overall performance to a

**Pirelli Société Générale**
(group managing director)

|

**Worldwide cable sector organization**
(cable sector general manager)

|

**Pirelli General plc**
(managing director)

|

4 operating divisions
(each with a divisional manager)

**Figure 1.3** *Management structure of the Pirelli cable sector and Pirelli General, 1988*

separate product-based cable sector general manager based in Basel (see Figure 1.3). As we shall see, the development of strategy within Pirelli General which led to the decision to build an experimental plant involved a complex interplay between managers at all levels, particularly those in the middle two.

## Pirelli General: History, Strategy and Structure

Pirelli General was established in 1914 as a joint venture of the Italian Pirelli organization and the UK General Electric Company (GEC) – thus Pirelli General. What brought the two companies together was the possibility of merging the engineering and cable product knowledge of Pirelli with the marketing expertise of GEC (GEC sold its interests to Pirelli in 1962). From the very beginning, the name Pirelli was a byword for high levels of technical and engineering expertise. In the early 1920s a Pirelli engineer, Luigi Emanueli, invented the oil-filled cable, a technical breakthrough which solved the problem of underground transmission of high-voltage electrical power and facilitated the creation of the National Electricity Grid in the UK. This highly profitable invention seemed to provide confirmation of a corporate strategy, business strategy and organization culture all driven by technical innovation and engineering expertise, from which sales and profits were expected to – and generally did – follow. This general approach allowed Pirelli General and the group's other national cable companies to grow slowly but consistently from the 1920s to the early 1980s in parallel with growth in the national electricity and telecommunications industries. The main stages in the history of Pirelli General are outlined in Table 1.1.

In 1984, the year in which the idea for the new experimental factory was first floated, the national production operations of Pirelli General were

**Table 1.1    *Key Dates in the History of Pirelli General 1909–1983***

| | |
|---|---|
| 1909 | Pirelli Ltd (UK) set up to distribute imported tyres, cables and footwear. |
| 1914 | Pirelli SpA (Italy) and the General Electric Company (UK) form a joint company, Pirelli General, and open their first cable factory on reclaimed land in Southampton. |
| 1924 | Oil-filled cable, developed in Milan by Pirelli engineer Luigi Emanueli, goes into production in Italy. |
| 1926 | Central Electricity Board set up to coordinate the national power system in the UK. In the following decade Pirelli General (PG) helps establish the National Grid, supplying oil-filled cable for underground sections in cities, while a new overhead lines department builds overhead sections. |
| 1928 | PG opens a new factory at Eastleigh, 8 km from Southampton, to manufacture power and telephone cables. The Southampton factory concentrates on producing general wiring cables. |
| 1940 | German bombers destroy Southampton factory on 30 November/1 December. Eastleigh site also damaged. |
| 1945–55 | After the War, the National Grid is overhauled, the Electricity Boards rebuild the cable network, and the Post Office re-equips the telephone cable network. A time of massive growth in the cable product market and sales. |
| 1959 | The Italian parent company resumes effective control of PG after 20 years under GEC following Italy's entry into the War in 1940. |

| | |
|---|---|
| 1961–3 | GEC reorganizes and sells its shareholding in Pirelli General to the Pirelli organization. The company decides on a root and branch modernization of the Southampton and Eastleigh factories. A number of senior UK managers are replaced by appointees from outside the company. |
| 1965 | A new oil-filled (super-tension) cable factory opens at Eastleigh. |
| 1967 | A new telephone cable factory is opened on a 75-acre site at Bishopstoke, 2 km from Eastleigh. |
| 1970 | Pirelli joins forces with Dunlop world-wide and becomes the largest tyre manufacturer in Europe. The mains cable factory at Eastleigh is modernized. |
| 1971 | PG puts a new organization structure in place, with strong corporate departments and four autonomous divisions: telecommunication cables, power cables (supertension, mains and industrial), general cables (rubber and plastic) and installation (later construction). Each has separate profit and operational targets appropriate to its particular product market. |
| 1971 | PG acquires Aberdare Cables Ltd for £2.75 million. The factory produces low voltage power cables. |
| 1979 | PG wins largest ever export order for 230 kilovolt cable in Singapore. |

**Table 1.1** *continued*

| 1980 | Pirelli General (40%) and GEC (60%) form RODCO to construct and operate a modern, continuous cast copper rolling wire factory in Skelmersdale. This leads to the closure of the Eastleigh Copper Rolling Mill in 1982. PG also acquires Meachers Transport Company, which transports finished cable to distribution depots throughout the country and copper rod from Lancashire to the Hampshire factories. | 1981 | Work begins to construct a new submarine cable factory in Southampton. The link between Pirelli and Dunlop is terminated, each returns shareholdings in the other's European operating companies to the groups to which they originally belonged. |
| --- | --- | --- | --- |
| 1981 | The Central Electricity Generating Board and Electricité de France sign a deal to interconnect both countries' national electricity supply systems by establishing a 46 kilometre, eight cable cross-channel link. PG wins the contract to supply the four submarine power cables for which the CEGB is responsible. | 1982 | PG wins an order worth £65 million to supply oil-filled cable and accessories to Kuwait. |
| | | 1983 | New submarine cable plant opened by Industry Minister Cecil Parkinson. |

*Source*: Pirelli General, 1989

organized on a multidivisional basis with four main product market divisions: *power cables*, the largest and most profitable group manufacturing high voltage cables for the electricity industry; *telecom cables*, for use in the telecommunications network; *special cables*, the smallest division, producing specialized cables such as those used in fire alarm systems; and finally – the focus of our study – *general wiring cables*. Each of the four divisions was run by a divisional manager, who was directly profit-accountable to the Pirelli General managing director (and through him to the PG board of directors, of which more in Chapter 2) and to cable sector management based in Basel. Each division and national company had an agreed five-year 'strategic plan', updated and revised annually in a divisional 'management plan' (known as the Manplan) containing agreed profit, sales and production targets.

## The general wiring product market

There are three main types of general wiring cables: building wires used for fixed wiring circuits in houses and buildings (the largest group in terms of sales); flexible cables used in domestic appliances such as cookers and kettles; and smaller industrial cables for use as fixed supply wires in smaller industrial installations.

General wiring cables are highly standardized 'commodity' products. There is little or no scope for product differentiation from one company to the next (a building wire is a building wire is a building wire). With little need for expensive research and development and relatively easy entry into the market-place, competition tends to be fierce, and although the volume of product sold is high, levels of profitability tend to be relatively low. The general wiring market is heavily dependent on the overall state of the national economy and the construction and consumer product markets. Another key feature of the product market is that the manufacturing companies do not control the supply chain to the consumer. Most of the users of building wires are (self-employed) electricians and electrical contractors, who tend to buy them in small batches as and when they need them rather than tying up money in keeping large stock holdings. Most purchases are made through the local branches of a small number of electrical wholesalers, who stock a wide range of products from different manufacturers. As a result, the manufacturers of general wires tend to be heavily dependent on the electrical wholesalers, which use their pivotal position in the supply chain to secure good discounts from the cable companies, with obvious implications for the latter's profit margins. The main characteristics of the general wiring product market are listed below:

- The products have to conform to national electrical safety standards and are largely interchangeable (they are sometimes referred to as a 'commodity' product).
- Manufacture does not require complex technology or expensive plant.
- Entry into the market on a small scale is relatively easy.
- Sales are highly dependent on the state of the housing and construction industries and of the economy as a whole.
- The manufacturers' immediate customers are not the state and the big utilities (as with power and telecommunication cables), nor the direct users, but a small number of electrical wholesalers who occupy a pivotal position in the supply chain.
- Competition is fierce and profit margins are low, both generally and compared with other cable product markets.

## Organization culture

As we have seen, profit, sales and production targets were an important part of the corporate and commercial framework within which each of the four divisional managers of Pirelli General (PG) had to operate. However, Pirelli General, and the Pirelli cable sector world-wide, was also rooted in an organization culture still infused with the engineering and technical values of its legendary founder, Giovanni Battista Pirelli. Everything PG produced

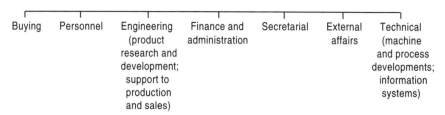

**Figure 1.4**    *Pirelli General corporate departments, 1985*

was not only well engineered, but often over-engineered. As one senior manager joked in 1990: 'our ideal is the Rolls-Royce cable'.

The corporate strategy and organization culture of Pirelli General in the early 1980s had two other distinctive features. The first, also strongly influenced by the character of its founder, was a long-term commitment to the core business of cable manufacture. This was the opposite of the short-termism which dominated much of British industry, finance and commerce at the time. Second, the company had a paternalistic and caring approach towards its staff. It only rarely made staff redundant and, when it did so, it tried to ensure that they had jobs to go to, or a good early retirement package, before they were dismissed. It aimed wherever possible to promote from within. It rarely disciplined employees who did not perform well. It had a well-funded sports and social club and an active retired members' association. This paternalism towards staff was reinforced by a kind of corporate paternalism of 'headquarters' departments. The large corporate engineering and technical departments, in particular, were centrally involved in all product and process innovations in the otherwise autonomous divisions (see Figure 1.4 for the structure of corporate departments in 1985). In contrast, marketing, sales and customer service functions were firmly devolved to the product divisions.

Engineering and technical excellence bolstered by strong corporate departments, long-termism, and a caring but insular paternalism towards staff: these were the foundations of the Pirelli General corporate strategy and organization culture. In contrast, some of the most widely accepted principles of contemporary business management – top-down financial controls, marketing, customer service targets – were not strongly in evidence in the business and operational strategy of the early 1980s.

## The Crises and Opportunities of the Early 1980s

By 1982–3 the poor profit performance of Pirelli General's general wiring division was beginning to give serious cause for concern among sector and group managers in Basel and Milan. So, when PG's managing director retired in early 1983, they decided to appoint in his place Alberto (Al) Marino, a senior American executive from their recently acquired US cable subsidiary. Marino came from a marketing – not engineering – background and sported a distinctive, most un-Pirelli-like Zapata moustache. His brief – and that of his deputy, Franco Ventrella, a long-serving Italian engineer – was to shake up PG, increase its general profitability, and above all tackle its languishing general wiring division.

The general wiring division not only personified all the elements of the traditional company culture, it was also at a critical point in its development. Profitability had been poor over the previous decade. It was in the middle of a price war which had resulted in large losses in 1982. Import penetration from low-wage countries such as Poland and Greece was increasing, particularly at the bottom end of the market (represented by a small but growing number of out-of-town builders' merchants and do-it-yourself stores). Perhaps most pressingly, the factory building and machinery were in need of major overhaul. The building had been erected in 1914 on reclaimed land and, although the factory had been redesigned and re-equipped in 1963, most of the machinery was approaching the end of its useful life. Operating, maintenance and overhead costs were relatively high, and the roof to the two-storey building, which also housed administrative staff from the general wiring division and all PG's major corporate departments, was in a poor state of repair.

## The five options

There were at least five policy options open to PG to tackle the problem:

1  Get out of the low-tech, low-profit general wiring market altogether and concentrate on the high-tech, high value-added cable product markets (submarine, telecom, supertension)
2  Stop producing general wires in the UK, but continue to supply them by importing from elsewhere
3  Acquire another general wiring company (or companies) and expand on a new site(s), with the option of eventually closing down the Southampton plant
4  Refurbish the Southampton factory and install new equipment where necessary
5  Build a new factory to manufacture the same or a new mix of general wiring cables

In 1983–4 neither of the first two options was considered seriously by the PG managing director nor by group managers in Milan and Basel (following usage within PG, 'Milan' is henceforth used as a collective term for group and sector managers, although some of them were based at the time in Basel). At first glance, Option 1 might have appeared to be the most logical solution for a company with a high-tech market profile and a Rolls-Royce (self-)image. But despite the profitability problems of the early 1980s, the four leading UK cable companies (BICC, Delta, AEI and Pirelli), with over 70 per cent of all UK manufacturers' sales, were all committed to continuing the manufacture of general wiring products.

In part this was a matter of tradition and self-esteem – a leading cable company needs to be represented in all the main product markets. But more importantly, all four companies believed that if they withdrew from the general wiring product market, their competitors would immediately swallow up their market share. This would then shift the focus of price competition to one of the other types of cable, which could be cross-subsidized through the increased profits from general wiring sales. They also assumed that, if they were unable to supply the electrical wholesalers with the complete range of cable types, they might lose out in favour of those who could. This is a classic example of a business strategy – here a commitment to continue in a particular product market – being determined not by objective, measurable criteria, but by a particular subjective interpretation of the market environment, together with a hypothetical judgement about the behaviour of their competitors and customers (if we do a, they might do b, with consequence c). This was the conventional wisdom in PG (and the other leading cable companies) in the early 1980s, and no one within PG sought to challenge it.

Option 2 – stopping production of general wires but continuing to supply them by buying them from elsewhere – raised its head briefly during board discussions in mid-1985, but was not considered in 1983–4. The most obvious way of achieving this would have been to import cable from one of the group's other general wiring factories in Europe or Latin America. However, this would have come up against the high tariff barriers which still existed at the time. In addition, continuing variations in the types of cable used in providing fixed wiring circuits for houses in different countries would have required the group's other factories to modify their machines and processes and extend the range of cable products they were manufacturing. Importing cable would have also involved increased transport costs and, more controversially, coordination and specialization of Pirelli cable production across Europe. In the early 1980s, coordination of cable production across the group's four European operating companies (France, Italy, Spain, UK) would have been a logistical, organizational and cultural impossibility. Each national company jealously guarded its own national product market, and it was not regarded as 'gentlemanly' for one national company to produce cable for the market of one of the others. It is a measure of 10 years of cultural and economic change that, by 1993, the European coordination of the company's general wiring operations had become a serious and increasingly compelling option (see Chapter 11). In 1983 it was not a practical possibility.

Option 3 – growing market share by acquisitions – was a strategy which had been pursued by PG's three major UK competitors in the 1960s and 1970s. However, it had already been decided at the highest level within the Pirelli group – PSG – to prioritize international acquisitions and new start-ups outside of Europe, where most of its cable business was still concentrated. Following the de-merger with Dunlop in 1981, the group was also looking to expand its tyre business through acquisitions, again mainly outside Europe. The upshot was that, in 1983–4, group management in Milan did not regard acquisitions as an appropriate way for PG to revitalize and grow its general wiring business.

This left Options 4 and 5. Option 4 – to refurbish the existing Southampton factory and install new equipment where necessary – was the traditional Pirelli approach to the manufacture of general wires, which had always been the Cinderella part of the business, with none of the major capital investments reserved for the high-tech areas such as submarine cables or optical fibres. However, Marino and Ventrella, PG's two most senior executives, rejected this option as postponing the resolution of fundamental problems – a production-limited and decaying factory building – while incurring significant expenditure. Wholesale refurbishment would also require production to be interrupted for substantial periods of time, thereby risking losing the increased market share which they had just gained during the price war.

## An Emerging Strategy

Option 5, to build a new factory, appealed to Marino. There were compelling reasons for staying in the general wiring market and for closing the Southampton factory. At the time, PG was comparatively 'cash rich' from its consistently good performance in the high-tech product divisions. Its corporate technical department had just completed the building of a new submarine cable factory, demonstrating its expertise in erecting and installing new factories. Finally, Marino and Ventrella had floated the idea of a new building wires factory with Milan, and, as we will see, it fitted in with Milan's rather different 'strategic' agenda. So, in the winter of 1983/4 Marino and Ventrella decided to incorporate the proposal to build a new general wiring factory in the first draft of their five-year strategic plan. It was the first formal step in the development of a strategy which was to lead to the construction of the most technically advanced cable manufacturing plant in the world.

However, all that had been done at this stage was to propose the construction of a new plant. Marino and Ventrella had consulted a small number of senior colleagues in the UK and Milan, and been encouraged to take things a stage further. But there was no analysis or back-up evidence to justify their proposal, no evaluation or costing of the various options, and certainly nothing like the detailed paperwork necessary to gain formal approval from sector management or the PG board of directors. To this end – with the support of Milan – they decided to carry out an internal time-limited feasibility study to explore, cost, and justify the proposal.

### Feasibility study

The man chosen to carry out the study was Barry Roberts. He was an obvious choice. He was the current divisional sales manager with responsibility for general wiring, knew the market and the product well, and had already been identified within the company as a high flier. This would be a chance for him to prove himself. He was given the task by Marino and Ventrella on the understanding that if the project went ahead, he would be both the project manager and the first divisional manager of the new factory. Gordon Peters, a member of the corporate technical department with previous experience as a production manager, was seconded full time to help him out.

The project report was completed in just over three months and submitted to Marino in June 1984. It was substantial – 38 pages of single-spaced text and 41 pages of supporting statistical material. It covered everything from product development, commercial (market) analysis, production, personnel, site location and finance. The report was so secret – particularly because of its final section which effectively recommended the closure of the Southampton factory – that Peters was not even allowed to see a copy, even though he had been involved in much of the preparatory work for it. The main findings of the report were:

1   A new dedicated factory unit would provide an 'acceptable' return on capital of around 15 per cent, although initial investment would need to be substantial.
2   The new unit would need to have all its critical activities (sales, accounts, personnel, warehouse as well as production and maintenance) on site to meet its goals of tight production schedules, reduced stocks and high product quality.
3   The preferred location was Aberdare in the Welsh valleys. PG already owned a site there and the area attracted regional development grants (see Chapter 3 below).
4   Most of the manufacturing technology was proven, but in some areas further investigation would be required (see Chapter 4).
5   Cost savings could be achieved by significant reductions in payroll, group working practices, and integrated computer systems covering manufacturing, accounts and distribution (see Chapters 4–7).
6   The level of working capital would be low due to faster turnover rates and reduced work-in-progress and finished goods stocks.
7   PG's relatively low share of the UK general wiring product market compared with its two main competitors, Delta and BICC (15/30/29 per cent respectively), meant that it should go for a less extensive distribution network and a more limited range of stock items. It should concentrate mainly on standardized building wires, which were particularly amenable to automated production.
8   Membership of a strong multinational group (Delta, for example, was mainly confined to the UK) with access to the group's R & D expertise would provide greater cost effectiveness in designing, choosing and installing new process technology.
9   Best market intelligence suggested that PG's UK market share would

remain constant at around 15 per cent and that total imports of general wiring cable into the UK would remain stable at around 10 per cent.

10   Underlying growth in the construction industry (and thus growth in the building wires market) from 1984–93 would be around 1 per cent per annum, subject to business cycle fluctuations of plus or minus 5 per cent; growth in the consumer durables market (the market for flexible general wiring cables) would be around 2 per cent per annum.

These were some of the major issues – but not always the actual outcomes – that were to dominate strategic and operational decision-making over the next 10 years. The feasibility study was the first attempt to justify a major new investment. But even at this early stage, the influence of Milan, and in particular the head of the cable sector world-wide, Arthur Stokes, was unmistakable. As Gordon Peters commented later:

> Arthur Stokes had the idea that we might do something very different in future, so this little team was set up to see what might be possible in terms of redesigning the factory, whether there was money to be made by setting it up in South Wales or elsewhere, to get some regional aid and to make the change . . . *The Feasibility Study was still limited in terms of its horizons, but it provided the answers that people seemed to want to know.* (Interview, 1 August 1990; emphasis added)

The report undoubtedly did provide the answers that Marino and Stokes wanted to hear. But the widening of horizons, and the development of a coherent and highly innovative strategic framework for the project, came predominantly from outside the UK, from sector management in Basel.

## The Multinational Dimension

To recap so far: the impetus for change came from many sources within the Pirelli group, particularly within Pirelli General itself. Profitability in PG's general wiring division had been poor for many years, and the factory building was in need of a major overhaul. The existing site was production-limited. Most of the machinery was coming to the end of its natural life and required substantial production and maintenance input. From 1979–81 manufacturing industry in the UK had suffered a major shakeout. In 1982–3 a bitter price war in the general wiring industry had resulted in a substantial year-end loss. In 1982 Milan had appointed a dynamic new managing director with a marketing background to shake up PG as a whole and its general wiring operations in particular.

And yet, despite all this, it was wider developments in the global general wiring product market, together with the vision of one individual, which placed the proposals for change in a truly strategic (long-term directional) framework. Organizational change is not simply the result of individuals reacting to problems. As Kanter has noted: 'People with solutions can go out for places to use them' (1983: 30). The major innovation we will be tracing in this book arose from the interplay between long-term structural problems

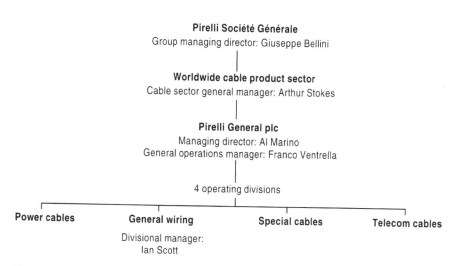

**Figure 1.5** *Management structure of the Pirelli cable sector and Pirelli General, 1984*

(plant, labour, profitability) in a particular global and national product market, galvanizing events (price war, the arrival of a dynamic new MD), new opportunities (resulting from advances in computer-based manufacturing systems and their application to standardized products), and the vision of one particular individual with a pivotal position in the organization structure. That individual was Arthur Stokes.

By 1984 Stokes had been with the Pirelli organization for 43 years. He had joined Pirelli General straight from school. After the Second World War he had taken an engineering degree at London University, gradually working his way up to become PG managing director in 1967. From 1971–82 he worked in Basel as head of the world wide cable operations of the Pirelli group (excluding the Italian operations, which at the time were part of a separate limited company). Then in 1982, following the de-merger of Pirelli and Dunlop, a new group management company was formed – Pirelli Société Générale (PSG) – and three new posts of 'sector general manager' were created for each of the group's main product divisions: cable, tyres and diversified products. The sector general managers had direct line responsibility for all the national operating companies in their particular product sector (including for the first time the Italian ones). They were directly accountable to the overall group managing director and also *ex officio* directors of the group management company. Arthur Stokes was the first cable sector general manager (see Figure 1.5 for an outline of the management structure in 1984 of the Pirelli cable sector worldwide and its UK arm, Pirelli General).

This new management structure cut right across the ownership structure of the company (in which Pirelli General appeared as a wholly-owned subsidiary of Pirelli (UK) and Société Internationale Pirelli, see Figure 1.2) and made it possible to develop for the first time a multinational commercial strategy for particular product markets worldwide. Stokes saw it as his first priority to set about trying to define the main areas in which 'sector' (as it came to be known) would influence and monitor the direction of the national operating cable companies. After nearly two years of internal discussions and visits to most of the cable sector affiliates, he brought together his thoughts in a 24-page discussion paper entitled *Evolution of Cable Sector Organization and Management*. He presented it to a conference of chief executives of the group's 14 national cable companies on 21 May 1984 in Milan.

In his paper Stokes argued that there would be two forms of multinational influence on national company operations: group level (PSG), and sector or product level. Group headquarters would play a major role in financial control and planning, and a limited role in purchasing and in personnel (executive development and career planning). Cable sector HQ would have overall responsibility for research and development and technical management (machine and process technology), and also establish performance targets and reporting procedures for the national cable companies. It was clear from the paper that technical management – which Stokes defined as the dissemination of best manufacturing practice and the development of new operating techniques – was his immediate priority. This fitted naturally with the group's and PG's corporate strategy and organization culture.

In his review of new operating techniques, Stokes concluded that national operating companies had failed to keep abreast of new developments in the manufacture of low-tech cable products, which were particularly amenable to the use of automated manufacturing technologies and computer-controlled production techniques. In this product area, too, competition within national markets was at its greatest, and cheap imports were beginning to make inroads into the lower end of nearly all the national markets in developed countries in which the company operated. The outcome had been consistently low profitability. He concluded: 'I believe that an imaginative, even experimental, approach is needed to improve our performance in this area'.

He then presented his own philosophy of change for the sector. This subsequently became known within PG as Stokes's 'tablets of stone'. The key passage from the report was:

> Analyse, in a questioning, critical spirit, every stage of the whole process from the acquisition of the raw materials to the encashment of the customers' payments for the products they have purchased, . . . aiming to devise new methods (or confirm existing ones) which will minimise total costs and average capital employed, so as to maximize the return on capital employed at whatever competitive prices can realistically be obtained for these so-called 'commodity' products.

In other words, all existing practices and techniques in the manufacture and sale of low-tech commodity products needed to be examined afresh, and either confirmed or rejected in the light of the evidence. However, rather than try this out in all of the group's 14 national cable companies, he argued that Pirelli General's proposed new factory, which would concentrate on the most standardized of all general wiring products, building wires, could act as both an experiment and a model for the whole cable sector.

Stokes believed it was crucial for sector management – and individual national companies – to find out whether investment in new computing technologies could offset the low labour cost advantages of competitors in Eastern Europe and the Far East, who were rapidly eating up market share in the global market-place. Stokes's hypothesis was that 'first world' companies such as Pirelli would only be able to compete in future by introducing new computer technology, both to reduce labour costs and to provide a quality product with the high degree of responsiveness to customer demand promised by computer-integrated manufacturing systems. The UK experiment would allow the dissemination of best practice throughout the national operating companies, and also show whether companies such as Pirelli could survive in future in highly standardized product markets. It was a proposal for 'transformational' rather than 'incremental' change (Mintzberg, 1978), using major new developments in computer technology to try and secure a continuing place in a highly competitive product market. If this approach did not prove viable, Pirelli might just as well cease competing in this particular market and concentrate on those product areas in which high levels of knowledge and skills gave them a competitive advantage.

Stokes admitted openly that neither sector nor national operating companies currently had the expertise for such a venture. He therefore proposed to establish a sector-level 'steering group' to coordinate work on the project. When it had completed its task, he proposed converting the steering group into a global industrial systems design team which would become a repository of knowledge and experience available to all the sector's operating companies world-wide. He also argued that the project would require new forms of collaboration with equipment suppliers and universities to solve complex technical and managerial problems requiring novel solutions.

The dovetailing of Stokes's multinational corporate strategy with Pirelli General's more nationally and operationally driven proposal for a new general wiring factory was not coincidental. Marino and Ventrella's 1983/4 strategic plan had already been thoroughly discussed and agreed in early 1984 by sector general manager Stokes and group managing director Bellini before it ever reached the Pirelli General board of directors. When it did reach the PG board for approval in the summer of 1985, four of the key board members were Marino, Ventrella, Stokes and Bellini (see Chapter 2). Stokes's paper to the May 1984 cable sector chief executives conference was a crucial stage in the development of a strategy which was evolving out of formal and informal discussions between senior managers at group, sector and national operating company levels.

It is also interesting to note that Stokes's paper was prepared before the

Roberts report had been completed, in other words, before it was possible to assess whether the proposed strategy was commercially and financially viable or not. Momentum was already building up by May 1984, and the most powerful figure in the structure of cable sector management was not only supporting the project, but to all intents and purposes driving it. The group managing director had also given his prior approval to the project, albeit with one proviso. Bellini wrote to Stokes on 16 May 1984 warning of the dangers of fragmentation if all the resources and expertise for such a major technical innovation were to be placed in the hands of one national operating company. Instead, he argued strongly in favour of giving the head of the sector technical department the authority and resources to 'guide' the development of new operating technologies in individual operating units. This was accepted by Stokes, with important consequences, as we shall see.

## Cometh the Time . . .

Stokes concluded his presentation of the proposed new project with the following words: 'the formation of a project team for such a project requires great care. The leader should be a person capable of imaginative and creative thinking, ideally the person who will eventually manage the new unit which may well become a teaching company for the cable sector.'

It was the job for which Barry Roberts had been originally selected. However, soon after he began work on the feasibility study, Marino and Ventrella decided that they needed him to run PG's power cables division. Over the previous decade this division had consistently been the major profit earner for the company with a turnover around three times that projected for the new building wires factory. In 1983, they had appointed a new divisional manager, but by April 1984 it became clear to them that he would need to be replaced. Roberts was the obvious candidate for the most prestigious operational management job in Pirelli General. It was a marvellous opportunity for him, and, as it proved, a stepping stone to becoming managing director of Pirelli's cable operations in Australia.

Stokes, Marino, Ventrella and PG Personnel Manager Richard Jones (of whom more later) now had to set about finding a replacement for Roberts as manager of the new project, by this time code-named 'Project X'. They concluded that there was no one within PG or the sector with the experience and capacity to manage it. At this point Ventrella came up with the name of a potential candidate. It was decided to put out feelers to see if he was interested.

On 12 May 1984, three men met for dinner in Romsey, a small country town in Hampshire in the South of England. The host was Al Marino. The go-between was Richard Jones. The guest for the evening was David Regan, a senior director of one of Pirelli's smaller UK competitors, Sterling Greengate. The next day Regan sent a *curriculum vitae* to Jones saying he was interested in taking things further. Jones wrote back rather formally on

23 May: 'Like yourself, we would wish to maintain our interest in the possibility of a career move on your part, and for this reason will be pleased to retain your papers on file in the event of a more positive development occurring'.

After this brief flurry of activity, nothing happened for seven weeks. Then out of the blue Jones telephoned Regan to ask whether he was still interested in a career move. Regan said he was, and they agreed to meet at 8.30 a.m. on 24 July in Jones's office on the first floor of the Southampton factory building. Regan still had no idea what the career move might be. Soon after his arrival in Jones's office, he was introduced to Ian Scott, divisional manager responsible for the general wiring factory. Regan, Scott and Jones chatted for about an hour, and it gradually emerged that Regan was being interviewed for a new position as assistant to Scott. However, Regan was to have no involvement at all in operational management. Instead, he was to be a one-person task force totally outside the official organization structure. His task would be to prepare the groundwork for investing in 'Project X', a state-of-the-art manufacturing plant to replace the factory currently located on the ground floor of the building in which he was being interviewed!

At the time of these events David Regan was 45 years old. He had 27 years' working experience behind him in five different UK cable companies. He had left school at 18 after A levels, later completing a four-year sandwich degree in polymer chemistry and progressing through a number of jobs in R&D, production and general management. On the way he had managed to attend a range of courses at some of the top business schools in Europe, including Cranfield, the London Business School, and INSEAD in Fontainebleau.

In 1980, he had been appointed managing director of the Sterling Cable Company, a wholly-owned subsidiary of the US-owned multinational defence and armaments manufacturer, Raytheon. Between 1980 and 1983, he had implemented large cost-reduction programmes, introduced new manufacturing systems, increased turnover, more than doubled market share, and raised Sterling's profit performance from under £100,000 to £1 million. In 1984 Raytheon amalgamated Sterling with Greengate, its other UK cable company, to form Sterling Greengate. Regan was rewarded with a promotion to the position of main board director with responsibility for product research and business development.

It was during this period (1980–4) that Regan first got to know PG's chief operations manager, Franco Ventrella. They had come to respect each other as competitors and as individuals, and Ventrella began to sense that Regan was not really happy at Sterling. Cable was not part of the core business of Raytheon, which had acquired its cable interests almost by chance as part of a general package of European acquisitions. Senior executives in Raytheon made it abundantly clear to Regan that they were not prepared to grow the cable business or make major capital investments. Regan became disenchanted and was clearly open to offers. [In 1989 Sterling Greengate was bought by BICC and its operation closed.]

By mid-1984 David Regan had reached a highly creative point in his career. He was energetic and highly knowledgeable about cable and the cable industry. He was experienced and successful in operational and general management and in managing advanced process technology. He was someone who got things done and was not afraid of being aggressive in the market-place. In this respect, as we will see, he was not a typical 'Pirelli man'. He was, it seemed, just what PG needed.

Just eight days before Arthur Stokes had presented his 'strategic' paper to the cable chief executives' conference in Milan, Regan had met Marino and Jones for dinner in a Romsey hotel to discuss the possibility of a career move to Pirelli. At the time he had no idea what the job might be. Two months later, he was offered, and accepted, the position which was to dominate his life over the next seven years.

## Summary

In this chapter we have outlined the strategy, structure and organization culture of Pirelli, the multinational group, and its UK cable-making arm, Pirelli General, until the early 1980s. The general strategy and organization culture were characterized by engineering and technical excellence, from which sales and profits were expected to – and generally did – flow, a long-term commitment to cable-making, and a caring but paternalistic attitude towards staff. In terms of strategy development, the ownership structure of the company was seen to be less important than the management structure, with its four main levels: group, cable sector, national operating company and division.

In 1982–3 the general wiring operation of Pirelli General – and to a certain extent general wiring manufacture more generally within the Pirelli group – had reached a critical point in its development. Profitability had been low for many years, cheap imports were gaining an increasing share at the bottom end of the market, and the factory building and equipment were in need of a major overhaul. Group management in Milan had recently appointed a new managing director of Pirelli General – an American with a marketing background – to shake up the company, particularly its general wiring division.

There were at least five policy options open to the UK company to deal with its general wiring operations, but the initial momentum was all in favour of the most radical one. In May 1984 the cable sector general manager, Arthur Stokes, announced his commitment to encourage an experiment in computer-integrated manufacturing in the general wiring operations of one national company. This dovetailed with a more operationally driven proposal by the two senior general managers of Pirelli General to build a new building wires plant rather than refurbish the existing one. These proposals were made before a systematic study could be completed to examine their viability. The internal candidate for the position of project manager for the new factory was suddenly no longer available, and so PG approached and persuaded David Regan, a senior manager from one of their smaller UK competitors, to take the post.

## Questions for Discussion

1  Discuss the usefulness of the distinctions between (a) corporate and business strategy and (b) 'planning' and 'emergent' views of strategy, in the light of the developments at Pirelli outlined in this chapter.

2  In what ways did the Pirelli General organization culture facilitate or inhibit strategic innovation and change in the early 1980s?

3  Discuss the significance of the feasibility study and report in the emergence of company strategy in 1984.

4  Which of the five approaches to the future of the UK general wiring product market would you have chosen in 1984, and why?

5  To what extent would you describe Pirelli General's and Pirelli sector management's proposals for a new general wiring factory as 'strategic'? For what reasons do you think having a strategy matters?

# 2

# Justifying Strategic Innovation

## Thematic Focus: Business Goals and Strategic Decision-Making

There is now a voluminous literature on the nature of business objectives and the various measures of commercial success (for a recent review see Pettigrew and Whipp, 1991: ch. 1). About the only point of agreement is the need to avoid focusing on just one measure of success. This is classically illustrated by the story of a nail-making factory in the former Soviet Union. The factory consistently exceeded its annual targets, which were based on the annual weight of nails produced. Together with his production engineers the factory director had worked out that the best way to maximize output was to produce nails one-third of a metre long. Nobody wanted to buy them, but he was always able to meet his target.

The influential neoclassical theory of the firm starts from the premise that the aim of business organizations and the entrepreneurs who manage them is to maximize profit. Organizational decision-making is about making rational choices between different known ways of maximizing profit. On this analysis, the key to successful decision-making is to express and model different choices quantitatively, usually in the form of financial comparators, thus allowing systematic comparison of the different options (see Whittington, 1993: 62–5). This approach gives a strong sense of legitimacy to strategic decision-making. Various quantitative techniques are used to help this kind of decision process, including different forms of investment appraisal (most strategic decisions require the investment of financial resources in buildings, equipment and people). As Whittington has argued:

> What these techniques share is an aspiration to approach strategic decisions in a structured rational manner that will finally produce a clear ranking of strategic options. The strategic option that promises the greatest net benefits, quantified in financial terms, should always be chosen. Rankings make strategic choice easy. In a sense technique obviates decision. (1993: 63)

The idea of rational decision-making based on quantified financial measures has some similarities to the idea of strategy as the product of rational planning about clearly defined problems outlined in Chapter 1.

The 'behavioural' critique of the neoclassical approach was developed, individually and collectively, by three American social scientists (see March and Simon, 1958; Simon, 1959; March, 1962; Cyert and March, 1963). In particular they argued that:

1 Neoclassical theory leaves open whether it is short-, medium- or long-term profit which is to be maximized.

2 Entrepreneurs may not decide in reality to maximize profit, but to restrict themselves to what they regard as a satisfactory return. This is Simon's celebrated distinction between 'maximizing' and 'satisficing' behaviour:

> If we seek to explain business behaviour in the terms of this [behavioural] theory, we must expect the firm's goals to be not maximizing profit, but attaining a certain level or rate of profit, holding a certain share of the market or certain level of sales. Firms would try to 'satisfice' rather than to maximize. (1959: 263)

3 Under conditions of imperfect competition and oligopoly (in contrast to the assumption of perfect competition in classical economics), profit maximization is an ambiguous objective. What is optimal for one firm may be suboptimal for another. In any event the achievement of profit maximization in one firm will depend on assumptions and expectations about the goals and actions of competitors, none of which can be predicted with certainty.

According to behavioural theories of the firm, alternative choices and strategies are not given, but need to be sought out and constructed by senior managers. No individual or even team of decision-makers can know everything about the internal and external environments of their firm. They therefore have to select particular problems and priorities and bring into play particular areas of information and knowledge (and leave out others) in order to make decisions. If there is a multiplicity of goals, and the internal and external environments of firms exhibit various degrees of uncertainty, then decision-making is about judgements and interpretations as much as about logical deductions. For the advocates of this approach, it describes more realistically what actually happens in organizations, whereas, in the words of Herbert Simon, 'the classical economic theory of markets with perfect competition and rational agents is deductive theory that requires almost no contact with empirical data once its assumptions are accepted' (1959: 254).

It is clear from the above that the questions of developing strategies and establishing objectives are closely linked to the nature of decision-making within firms. Following Baldridge (1971), there are three main models of organizational decision-making (see Table 2.1). In reality, it is rare for any of these models to be found in pure form. While power is often a major factor in organizational decision-making (the political model), the ability to apply it in particular cases is often dependent on the position of individuals and groups within the structure of the firm (the bureaucratic model). At the same time, the views and expertise of professionals may also be instrumental in coming to a particular decision (the professional model).

The political model is particularly useful when analysing strategic

**Table 2.1** *Models of organizational decision-making*

| Model | Nature of decision-making |
| --- | --- |
| 1 Bureaucratic | Organization seen as hierarchical structure with rational universalistic criteria and formalized rules and procedures; decision-making is about the rational optimization of clearly defined, unitary goals. |
| 2 Professional/ collegial | Organization as non-hierarchical community of professionals or colleagues with rational universalistic criteria and informal rules; decision arises out of inter-personal consultations and debates. |
| 3 Political | Organization seen as coalition of divergent perceptions, interests and goals; decision outcomes depend on the ability of individuals and groups, particularly the 'dominant coalition' (Cyert and March, 1963), to apply power and influence in particular contexts and on particular issues. |

decision-making and trying to understand processes through which dominant coalitions seek to achieve particular objectives – what the US political scientist R. A. Dahl has called 'the mobilization of bias'. For Dahl, the base of an actor's power consists of 'all the resources, opportunities, acts and objects that he [or she] can exploit in order to affect the behaviour of another' (Dahl, 1957: 203). The political model also helps our understanding of those not uncommon processes of organizational decision-making in which forecasts or rationales are developed to justify desired outcomes (for an early cited example, see Cyert et al., 1958) rather than the other way round, as one would expect under rational-bureaucratic models.

In looking at the relative merits of different models of decision-making, it is also important to consider Simon's distinction between routine, repetitive, programmed decisions and novel, non-routine, non-programmed ones. While many decisions to be examined in the following chapters will be adaptations or minor modifications to strategy, we are concerned in this chapter with the justification of a strategic investment decision involving substantial degrees of technical – and, as it proved, many other forms of – innovation. Under such circumstances of uncertainty and risk, it is likely that there will be greater scope for interpretation, judgement and mobilization of bias than in smaller more routine decisions. As Pettigrew has noted (1973: 21): 'political behaviour is likely to be a special feature of large-scale innovative decisions'.

The third main theme of this chapter is the idea of a 'technology-driven strategy'. The concept was first mooted in the *Harvard Business Review* in 1981 and soon came to signify a full-blown critique of US corporate governance (Abernathy et al., 1981, 1983; see also Hayes and Abernathy, 1980) in which US business leaders were accused of failing to meet the growing competitive challenge from Japan and other Far East countries.

The core of the argument was that US business strategies were too fixated on short-term financial objectives and measures – such as annual return on capital invested – and unprepared to engage in longer-term product and market innovation. In contrast, Japanese corporate and business strategies appeared to be much less guided by financial rationality and more by vision, faith, and what Whittington has called 'grand sweeping statements of ambition' (1993: 69). Pettigrew and Whipp have summarized neatly the core of the debate:

> The cult of quantification, allied to the primacy of financial and legal skills, left little space or influence for technological ideas. By contrast, the Harvard group saw overseas competitors' growing superiority arising directly from their strengths in technological innovation . . . they saw a series of revolutionary changes emanating most especially from the ability of Asian companies to link high levels of innovation in not only product but production and management systems . . . [By] combining technological and market changes the problems of rigid product designs and the dead weight of elaborate production processes could be overcome. (1991: 17, 18)

Subsequent criticisms of technology-driven strategies, fuelled by the apparent failure of some large-scale technical experiments (see Chapter 4), have argued against an over-reliance on technology and large-scale experimental projects as a means of turning round uncompetitive companies. They have also reaffirmed the crucial importance of financial controls and appropriate commercial strategies in managing successful companies (see Pettigrew and Whipp, 1991: 18–19; Grinyer et al., 1988). Nevertheless, the concept of technology-driven strategy did draw attention to the way in which investment in new computing technologies had enhanced the competitiveness of Japanese companies. It also highlighted their ability, in the medium and long term, to help reduce costs and adapt flexibly to changing market requirements. In other words, technology-driven strategies were intended to achieve exactly the kind of financial and commercial advantages that subsequent critics accused them of neglecting.

---

# Preparing the Case for Strategic Innovation

## The technical project team

David Regan, a cable-maker all his working life, joined Pirelli General at the age of 45 because he was not convinced that his current company was committed to the cable industry long term. He had known Pirelli for many years as a competitor, and was impressed by its high reputation in manufacture and production: 'everyone believes that if Pirelli does something, it does it well', he recalled later. As if to prove it, the company had just completed building a world-class, state-of-the-art submarine cables plant in Southampton.

Regan started working full time for PG in January 1985. He already knew that the managing director (Marino) and the general operations manager (Ventrella) were about to move to jobs in the group outside Britain, and there would inevitably be some uncertainty at the top. However, he was convinced that it had already been decided to build a new general wiring factory and that his main task as project manager was to convert the idea into reality.

It took only a few days for him to learn that the formal decision to go ahead had not been taken. He also soon found out that he was in a highly anomalous position in the company. His work was to be top secret (code-named Project X) and only to be discussed with a small number of chosen individuals. He was given a room in company HQ in Southampton, but the first floor of the building was such a maze of corridors and blind alleys that he was assured that few people would know of his existence. This was important, since the implementation of Project X almost certainly meant the closure of the Southampton factory. Special locks were put on his door and, unusually for PG, the door had to be locked at all times when he wasn't there. He appeared on no organization chart and no mailing or circulation lists. He had no job description, kept no minutes of meetings, and only rarely sent memos.

To judge by the account so far presented in this chapter, it may appear as if David Regan almost had to start from scratch when he joined PG in January 1985. But, while no formal decision had been taken to proceed with the new project, he soon discovered that there was already substantial momentum behind it. The feasibility study by Barry Roberts had prepared some of the initial ground, particularly in terms of market research and the decision to specialize in the manufacture of building wires, the most standardized type of general wiring cable. Perhaps most importantly, in May 1984 sector general manager Arthur Stokes had publicly expressed his commitment to the building of an experimental, state-of-the-art general wiring factory in front of the CEOs of the group's 14 cable operating companies world-wide. In September 1984 Stokes had also established an industrial automation steering group in Milan to begin a detailed investigation of the various technical options for the new factory. From January to May 1985, a small team of four people from this group – David Regan, Gordon Peters from the PG technical department and two members of the sector technical department with substantial experience of machine innovation and plant installations – formed the core of what might be called the 'technical project team'. This team did much of the detailed work to prepare the formal case which was eventually put before the PG board of directors in June 1985.

The project team was allocated rooms next to Regan in the PG headquarters building, and like Regan appeared on no organization chart. The aim of this cloak and dagger approach was to keep everything secret until the final board decision was made. At the same time, it did create an atmosphere of excitement and innovation. Regan and his three colleagues were thus able to give full rein to their imagination in the early days of the project, untrammelled by traditional company rules and procedures. They also established a collegial, team approach to policy development which was to become one of the hallmarks of the project.

### The need for a financial justification

However, there was one crucial element still missing if the project was to stand any chance of securing approval from the PG board of directors. It would clearly require a costing and detailed financial justification. This had already been recognized at the first meeting of the steering group in Milan in September 1984. Immediately following this meeting PG accounts manager, George Sherfield, had been asked by Ian Scott to second a senior member of his staff to make a financial input to the work of the steering group. This was agreed, and Paul Bamford was seconded to the position. He was an experienced 'Pirelli man', having joined the company in 1968 and worked for 10 years as PG's chief accountant before going to the group's Canadian operating company as director and vice-president (finance and corporate planning). He had rejoined PG as corporate development manager in July 1984, and this was his first major task on his return.

Bamford set to work from day 1 with tenacity and commitment. Although still formally responsible to Sherfield and through him to the sector level steering group, he effectively worked full time for David Regan from January to September 1985. During this time he came to play a number of different roles. First and foremost he was the financial adviser to the project. In this role he continually had to press Regan and his colleagues on the technical project team to specify clearly the machines, buildings, finished goods stock levels, work-in-progress, materials stocks, and numbers of staff they would need so that he could work out the costings of the various items. This was extremely difficult given that many of the machines and processes were highly experimental and currently not available for purchase. At this stage, too, it was only possible to make guesstimates of the number and types of staff required.

## Generating Consensus

### The involvement of the new PG managing director

In February 1985, David Regan was formally requested by sector general manager Arthur Stokes to give an update on progress with the project. In particular he was asked to present a detailed review of any problems not yet resolved and to outline his ideas on how the automated manufacturing equipment would be integrated into the wider commercial, financial and management information systems of the new plant.

In the weeks leading up to the March meeting, another key player in the strategic development of the project began to make his presence felt. This was Giuseppe Mancini, Marino's replacement as managing director of Pirelli General. Mancini had worked for Pirelli all his life. He had wide experience of general management, production management and plant installations. This included a spell in Britain as works manager and divisional manager of the telecommunication cables division from 1966 to 1975. In 1975 he had returned to Italy as director and general manager of the telecommunications group within the Italian operating company, and no one had envisaged him moving again before his retirement. Marino had taken up his job as PG managing director at the beginning of 1983, and had been expected to stay in the post for around four or five years. However, he departed from the UK at the end of 1984 for domestic reasons – his family had never moved from the States and he had found the commuting very tiring – and in his place group management appointed the tried and trusted Mancini.

The contrast in background and style between Mancini, the conventional, respected, time-served Pirelli engineer, and Marino, the flamboyant American marketing executive appointed in 1983 after only four years' previous experience in the company, could not have been greater. However, both were equally committed to the project. For Marino it represented an opportunity to break out of the cycle of decline, inefficiency and neglect that had typified the general wiring division since the 1970s. The fact that sector managers were in favour was not an end in itself, but a means to an end. For Mancini, the building of the new automated factory in Britain was a publicly-expressed goal of sector general management in Milan and that was enough for him. He wasted little time after his arrival before reinforcing this message. Just seven weeks after taking over, he wrote to his main corporate managers and to Regan and Bamford reminding them of the wider context of the project, and drawing particular attention to the views of his immediate boss, Arthur Stokes:

> At the cable sector CEOs' meeting in May 1984, Mr Stokes formulated some basic principles that should be applied in trying to study and implement an industrial automation project on business lines. As we are [near to completing Phase 1 of the project], I feel it is appropriate that full consideration should be given to the principles and approach suggested by Mr Stokes.

To this end, he attached to the memo a photocopy of the key sections of Stokes's presentation in Milan in May 1984, including the 'tablets of stone' reproduced in Chapter 1. From this moment to his retirement in 1989, Mancini ensured that the project was implemented as much as possible in conformity with the requirements of cable sector management in Milan.

## The update presentation March 1985

The 'update' presentation took place on 14 March 1985 in the board room at Pirelli General HQ in Southampton. It was attended by the five most influential members of cable sector management: Stokes (director and sector general manager), Guido Angeli (deputy sector GM and GM-designate), and the heads of the sector finance, R & D and technical departments – together with two specialists in computer systems and software from the group's Italian-based optoelectronics subsidiary. The meeting was chaired by Pirelli General MD Giuseppe Mancini, and attended by the heads of five PG corporate departments – accounts, engineering, information systems/data processing, personnel, technical – together with accountant Paul Bamford and PG's chief engineer.

The meeting concentrated on two main areas, technical/engineering matters and project costs. Issues such as product mix, commercial strategy, customer service, and personnel were either not discussed or appeared as a function of technology or finance (for example, 'headcount' reductions were included as part of the financial calculations). This narrow focus was partly the result of the limited time available. It was also based on a judgement about the most pressing unresolved problems. However, it reflected, too, the traditional Pirelli engineering culture, the largely technical membership of Regan's core project team, and the background of those attending the presentation (12 of the 16 people present came from an engineering background). The whole morning session lasting three hours was spent discussing technical matters, while only one hour in the afternoon was devoted to project costs. These were broken down into three headings: location, cost, and timetable.

By all accounts Regan's 90-minute morning 'sales pitch' on behalf of the project was a *tour de force*. He used the whole gamut of visual aids to range widely over all aspects of computer-integrated manufacturing (CIM). (CIM was at this time the newest 'buzz word' in engineering circles, promising the computer control and integration of all aspects of the manufacturing process, and indeed many areas of pre- and post-production, in one fully automated process: see Chapter 4 for more details.) Everything he said assumed that the factory would be an 'experimental' or 'model' project for the sector as a whole, as Stokes had intended. The initial costings for the new equipment (£17 million) and buildings (£3 million) were based on guesstimates by Regan and Bamford for what were in many cases extremely innovative machines and automated handling and control systems (see Chapter 4). A number of different sites for the new plant were also considered and costed out (see Chapter 3).

The meeting had three main outcomes. First, sector representatives, led by Stokes, expressed strong support for the proposal as presented. However, they felt the projected cost of £20 million and the five-year payback period were much too high to be approved by the PG board (of which Stokes was vice-chairman, see below). Regan and Bamford were therefore charged with preparing a number of options for the board with differing degrees of automation and cost, all of which were required to have a shorter payback period. Second, it was agreed that revised calculations needed to be prepared on the relative profitability of the various options. On the advice of George Sherfield, head of the PG accounts department and finance director of the company, profitability was to be expressed in terms of specific technical accounting measures over a 10-year period (of which more below). Third, Stokes agreed to put some as yet unspecified money from the sector research and development budget into those aspects of the project which were not 'economical' in the strictest sense, but served the wider group interest as part of a world-wide experiment in new operating techniques.

This meeting effectively established the general framework for the future development of the project. Sector representatives were clearly in favour of the general strategy of setting up a state-of-the-art computer-integrated general wiring factory in the UK and would clearly support it in the coming board discussions. However, neither the projected capital costs nor the payback period would be likely to be acceptable to the board and both would need to be scaled down for the project to go ahead. Mancini, Regan and Bamford were given two months to prepare the revised justification in time for the next quarterly board meeting in June. The hope was that this would be a relative formality. It wasn't!

# Conflicting Objectives and Investment Justifications

Between March and June 1985 a fundamental conflict developed between two different approaches to strategic investment decisions. The conflict mirrored the different horizons and perspectives of multinational managers – interested in the development of a global strategy for a particular product market – and senior managers of one national operating company concerned with the profitability of their particular wholly-owned subsidiary. It was also overlaid by a debate about the role of the finance function in strategic decisions and the appropriate criteria by which to justify a large-scale, high-risk capital investment in new technology. It is therefore worth looking briefly at the debate about cost justifications for computer-integrated manufacturing systems before we look in detail at the actual events at PG.

## Cost justifications for CIM systems

Some of the greatest benefits of CIM systems – such as savings on labour and on stocks of raw materials, work-in-progress, and finished goods – can be quantified using traditional cost accounting techniques. However, from the mid-1980s onwards leading engineering, accounting and management journals in the UK and North America became increasingly critical of the relevance of such techniques in assessing the wider costs and benefits of computerization and advanced manufacturing systems (see Bonsack, 1987; Meredith and Hill, 1987; Storm and Sullivan, 1990). Indeed, techniques such as the 'payback period' and 'return on investment' criteria were variously described as 'a primary obstacle to companies' adoption of CIM'

(Rohan, 1987), 'a troublesome barrier to automation' (Bolland and Goodwin, 1988), or even as 'the overwhelming factor keeping . . . companies from fully adopting CIM' (Patterson, 1987). In these articles reference was continually made to the long-term benefits of CIM (see Meredith and Hill, 1987) and to the fact that successful Japanese companies nearly always downplayed the importance of short-term profits and horizons, and financial measures more generally, in making strategic investment decisions (see Huang and Sakurai, 1990; Whittington, 1993: 69–71).

In contrast to these views, some leading commentators argued that, while many traditional methods of cost accounting might be inappropriate, the longer-term benefits of automation could be adequately assessed by the use of one particular technique, the discounted cash flow (DCF) method (see Bolland and Goodwin, 1988). As this method played a major role in the PG case, it is worth looking at it in a little more detail.

The DCF method is a financial measure of profitability over a particular period (say 5, 10, or 20 years). It allows a quantification of the relative rate of return of various capital investment options. When making an investment, it is important to calculate not only *the amounts* which need to be spent, but *when* they will be spent in the life-cycle of the project. The same is true for projected income from the investment. Once a net total inflow and outflow of funds is prepared for each year, say over a 10-year cycle, a discount is applied to the net position in each year, i.e. £100 income in four years' time is worth £100 minus a discount represented by a calculation for depreciation in the value of money. Once this is done, the discounted figures for each year are added up and the result is a single figure, a 'net present value' or DCF return on capital (see Bolland and Goodwin, 1988, for more details). A DCF return is, in other words, a particular form of calculating the return on capital invested in a particular project, based on the total cash inflows and outflows over the equipment's projected lifetime. Like all such projections, these calculations are an inexact science, since judgements have to be made about the discount rate to be used (including assumptions about inflation and interest rates) and the amount of income generated from sales, all of which may vary substantially over a 10-year period.

Only a tiny minority of critics has gone as far as to advocate the complete abandonment of financial justifications in evaluating large-scale investments such as new manufacturing systems (see Drury, 1990). At the very least such justifications are a useful discipline. They provide the appearance, if not always the substance, of rationality which is so essential to the legitimacy of any investment decision (see Whittington, 1993: 62–71). However, it is now generally accepted that many of the benefits of automated systems are not only long term, but intangible. The intangible benefits include their capacity to allow (a) continuous improvement of operating procedures and product quality, (b) continuous technical innovation and (c) ever greater flexibility and responsiveness to customer orders. Most commentators also emphasize the importance of the 'faith' or 'vision' needed when deciding to introduce

CIM or other advanced technological systems. All underline the importance of justifying such investments in terms of the overall strategic objectives of the organization. The conflict over financial justifications is neatly summarized by Storm and Sullivan:

> CIM investments are difficult to justify because traditional measurements overemphasize short-term quantitative factors and undervalue the long-term qualitative factors that create marketplace advantages. CIM must be viewed as a strategic commitment and justified by its contribution to long-term corporate objectives. (1990: 40)

### The role of the PG accounts manager and the board of directors

The central figures in the debate about investment justifications at Pirelli General were Arthur Stokes and George Sherfield. Until the meeting of 14 March 1985, Sherfield had played virtually no role in Project X. A senior member of his department, Paul Bamford, had been a full member of the project team since November 1984. However, Bamford had effectively been working full time for David Regan since 1 January 1985 and become more closely identified with the project team than with the accounts department.

Sherfield's role in the process of strategy development was influenced by two quite separate factors. First, as head of the accounts department and also finance director of Pirelli General, he was a full executive member of the PG board of directors which was required formally to approve the decision to go ahead with the investment. He thus had considerable formal positional power in the decision-making process. Second, he had only recently joined Pirelli General from GEC (in 1983) and was strongly imbued with the latter's business philosophy. He believed GEC's concentration on short-term 'bottom line' objectives, its strong corporate financial controls and reluctance to spend profits on large-scale investment in new plant and machinery, had all contributed to making it into probably the most commercially successful UK engineering company over the previous 20 years. In many ways Pirelli General was the complete opposite of GEC: a family company committed to the long term, engineering- rather than finance-led, with a relatively poor record on profit growth and a paternalistic caring approach towards its staff.

During the 14 March update meeting Sherfield had become increasingly concerned about the financial case (or lack of it) for the project. In the following weeks he took a much greater interest in the costings of the various options which were to be presented to the board. He also asked Bamford to do a detailed calculation of the costs and financial return of refurbishing and upgrading the Southampton site, an option previously rejected by Stokes and Marino in 1984 (see Chapter 1). In many ways this option represented a typical GEC approach. The feasibility study of June 1984 had included a thumbnail sketch of each of PG's three main competitors in the general wiring market, and had concluded that AEI (a GEC subsidiary) had adopted 'a "shoe-string" approach to most activities' with 'little indication of investment in either manufacture or distribution'. For Sherfield, such a

| Chairman | Lord Westgate<br>ex-Conservative Party Chairman and ex-Chancellor<br>of the Exchequer |
|---|---|
| Vice-Chairman | Arthur Stokes<br>General manager, Pirelli Société Générale cable sector |
| Executive directors | Giuseppe Mancini<br>Managing director, Pirelli General |
| | George Sherfield<br>Finance and administration manager, Pirelli General |
| | Raymond Barlow<br>Company secretary, Pirelli General |
| Nominees of parent company | Giuseppe Bellini<br>Chairman and managing director, Pirelli Société Générale |
| | Guido Angeli<br>Deputy general manager, Pirelli Société Générale cable sector |
| | Enrico Urbani<br>Director, Pirelli SpA |
| Non-executive directors | Sir David Marshall<br>Chairman, Civil Aviation Authority |
| | Peter Keep<br>Director, Morgan Grenfell (merchant bank) |
| | Martin Hofmann<br>Director, Lazard Brothers (merchant bank) |

**Figure 2.1** *The company board of Pirelli General, 1985*

shoe-string approach was a real option for PG in 1985 and he worked with Bamford to prepare a detailed financial case for it.

Sherfield had also become convinced at the March meeting that Project X was being driven too strongly by the engineering concerns of sector – and what he believed was its desire to experiment with the latest technology – and that cost and profitability should be much more prominent as criteria in strategic decision-making. This view of the nature of investment justifications was complemented by (some would later argue, conveniently concealed behind) a particular view of the role of finance director in board decision-making. Sherfield believed it was his task to ensure that the board was presented with a full range of alternatives from which it could then choose. In March 1985, it appeared as if the board might only be presented with a choice between different levels of automation, whereas he believed a refurbished Southampton factory option should also be put.

The composition of the Pirelli General board of directors, which alone had the formal powers to approve such a large-scale investment decision, was a crucial factor at this stage in strategy development. Of its 10 members – excluding the chairman – 7 were full-time Pirelli employees, but only 3 worked for Pirelli General, the rest being members of multinational group management based in Basel or Milan (see Figure 2.1). In addition, the most important Pirelli General executive director, MD Giuseppe Mancini, was a recent Milan appointment who had already expressed publicly his commitment to implementing sector's declared objective to set up an experimental building wires factory in the UK. However, there were also three non-executive directors on the board, one a leading industrialist, the other two merchant bankers from firms in the City of London. This meant that financial justifications for major investment proposals were likely to be subjected to close and informed scrutiny. For all these reasons, Sherfield fought hard for a fully-costed Southampton refurbishment alternative to be placed before the board in June.

### The build-up to the board meeting, June 1985

In the build-up to the board meeting, two parallel processes were taking place. Regan was working with Bamford to prepare a series of revised versions and costings of the original proposal based on three different levels of automation for the new factory. These options – now no longer headed Project X (PG terminology) or industrial automation project (sector terminology), but computer-integrated-manufacturing project – were sent to Stokes and his sector colleagues for information. At the same time, Bamford was also preparing calculations on the Southampton refurbishment alternative.

During this period the criterion of discounted cash flow return weighed heavily in discussions of the various options. By the end of March 1985, Bamford had prepared the initial calculations of DCF returns for what had emerged as the four main options:

1   Project X at Aberdare with a capital investment of £16 million (reduced from the original £20 million)
2   Project X at Aberdare with capital investment of £13.5 million (£14.8 million with £1.3 million deferred)
3   Option 1, but deferred for three years with some interim expenditure to keep Southampton going
4   A refurbishment of the existing Southampton factory, with an estimated cost of £7.5 million.

The first three options were based on a 'payback' period of four years (reduced from the five at the March update meeting). The results of the DCF returns were respectively: 19, 23, 16 and 46 per cent. In other words, the DCF method showed that the most profitable option, calculated in terms of the relative rate of return over 10 years with a payback period of four years, was the refurbishment option. All the options to invest substantial sums in new technology were, on these criteria, much less profitable.

Calculations for the first three options were sent by David Reagan on 15 April to sector management for consideration, with the comment: 'the options all provide for a CIM plant which is not significantly diluted from cable sector management requirements'. All represented a reduction in total capital outlay as requested by sector, and option 2 was made to look particularly attractive, reduced to £13.5 million by using refurbished machines in some areas and by deferring other expenditure to an unspecified future date. At this point, sector was not sent details of the fourth option, although these were made available to Mancini and Regan.

The weeks leading up to the board meeting involved a hectic backwards and forwards between Mancini, Sherfield, Regan and Bamford. At this point, the position of the main protagonists within the formal decision-making structure of the company was crucial. Mancini and Sherfield were members of the board. In addition, Sherfield was PG's head of accounts. While this position did not require him formally to approve budgets or capital expenditure (a reflection of Pirelli's engineering-led tradition), his department was required to 'keep the score' and cost large-scale proposals. More importantly, given the presence of the non-executive directors, it would have meant instant rejection for a major investment proposal to be put to the board by the managing director when his finance director (who was also on the board) was not happy with it. Against this background, Sherfield and Mancini finally agreed the form of the submission to the board, which represented a compromise between their two positions.

## The board meeting, June 1985

The PG board was eventually presented with a document outlining five options: a 'recommended option', and four alternatives. The recommended option involved high levels of technical innovation, promising improved lead times, reduced work-in-progress, improved process control, new efficient plant layout, and the linking of previously separate machine processes. It was an investment justification based on the reduced production costs expected to flow from a high level of technical innovation. It showed a DCF return of 21 per cent and an initial capital outlay of £13.5 million. This had been achieved by scaling down many of the initial costs and postponing some of the initial capital outlay to arrive at the kind of overall figure required by Stokes. In carefully chosen words, this proposal was held 'to provide the best economic return in line with the terms of reference of the project'. In other words, economic return was an important consideration, but this needed to be seen in the light of the particular 'terms of reference' of the project, that is, the requirement to conduct a state-of-the-art experiment on behalf of the multinational group.

Having argued a detailed – mainly technical – case for this option, the document then went on to outline four possible alternatives for comparative purposes:

1  The project as presented, but with a capital outlay of £16 million, a DCF return of 20 per cent, and a payback period of four and a half years.
2  The project deferred for three years – DCF return of 19 per cent.
3  The project as presented, but eliminating new equipment which did not

show a direct return (such as automatic materials handling equipment) and utilizing much more of the existing Southampton plant. This had a projected DCF return of 29 per cent.
4  The refurbished Southampton factory. This was a low-tech mix and match approach, involving a capital outlay of £7.5 million over a 10-year period, £3 million of which would not be spent until 1989 and beyond. It showed a DCF return of 48 per cent.

The board had a lengthy discussion of the various proposals. Alternative 1 was a non-starter. Although it was the option that sector representatives really favoured (the so-called 'all singing, all dancing' factory), they had already decided at the March meeting that the capital outlay required would be too great to get the approval of the board. It was there as a stalking horse and certainly made the recommended proposal (with reduced capital outlay and slightly higher DCF return) look more attractive. Alternative 2 had the advantage of the 'me too' approach to innovation, the view that you should never be the first to innovate but wait for someone else to take all the risks and learn from their mistakes. However, it would almost certainly have involved a reduction in development grants (see Chapter 3) and required at least some interim expenditure in Southampton to keep the factory going for three years. For the advocates of Project X, it would have meant the end of the project, as all the momentum would have been lost. The decisive argument against this proposal, however, was the DCF return of 19 per cent, the lowest for any of the options. (Given that all the DCF returns were estimates, they could only be regarded as accurate within a 'noise level' of plus or minus 5 per cent. In this sense, the difference between 19, 20 and 21 per cent was not statistically significant.)

Alternative 3 was not unattractive from the national operating company perspective of Pirelli General. For sector representatives it was unaccept-able. It would have meant the end of their wider plan for a group-wide experimental testing ground in the use of new technology. Alternative 4 was Sherfield's preferred solution, but was opposed strongly by sector rep-resentatives as piecemeal, unimaginative and short-sighted. There was even an alternative 5 proposed by one of the non-executive directors, who suggested that PG should still sell building wires in the UK, but should import them from the Far East (e.g. Taiwan) at low prices and market them under the Pirelli name. No one else supported this option, above all on grounds of product quality.

The outcome of the long discussion was a compromise. The recom-mended option was approved 'in principle', but the release of funds was 'referred to the next meeting' pending more detailed calculations and information on various matters such as plant utilization and market share. To paraphrase one of the board members: 'the concept is fine, but we don't like the numbers'. This is an interesting example of the *de facto* powers of a board of directors, with its mixture of senior executive managers and non-executive directors (the latter once described rather harshly by ex-chief executive of Lonrho, Tiny Rowland, as 'decorations on a Christmas tree'). In effect, the role of the PG board was to receive proposals, plans and budgets prepared by the managing director and his executive team. Since it was a multinational company, these had already been cleared, and in this case championed, by senior group managers, who were also strongly represented on the board in their own right. In practice the task of the board, and particularly its non-executive directors, was not to tell executive managers what to do, but to advise and comment on plans and projects and point out their risks, weaknesses and strengths. However, the fact that the proposal had to be put to the board, on which there were three non-executive directors, two from finance and banking backgrounds, forced executive managers to prepare a detailed and plausible case in financial terms. In the end, the biggest sanction of the board was to delay the proposal and refer it back for further work. And this is what happened in June 1985.

# The Outcome: A Technology-Driven Strategy?

### *Re-costing the two main options*

Stokes and Mancini were distinctly unhappy at the outcome. It is no secret that some choice words were exchanged in the immediate aftermath of the board meeting. Predictably, too, Sherfield was not the most popular man in the corridors of Pirelli General HQ. However, although approval in principle represented a small delay, it also meant that in practice the experimental project was now unstoppable. All that was needed were revised DCF calculations which would make the CIM option more attractive financially. In anticipation of board approval, a meeting of the industrial automation steering group (now called the 'CIM automation project' group) had already been arranged by Mancini for 17 June. At the meeting he reported on the board decision and began to assign detailed tasks to specific individuals (for details see Chapter 4).

Parallel to these developments on the technical side, Bamford set about preparing revised justifications and gathering more detailed information about the output and market share implications of the two key proposals – the recommended option and the Southampton refurbishment option. With the admonitions of Mancini ringing in his ears, Sherfield became convinced that the original calculations for the refurbishment option – which had been largely a 'desk' exercise prepared for illustrative purposes – were too optimistic. Repairs to the floor, the roof and the equipment in the Southampton factory were likely to be much more costly and immediate than he had initially envisaged. The maximum volume of cable which could be produced in the Southampton factory was also severely limited.

On the other hand, the 'recommended' option had always assumed roughly the same output and market share as was currently being achieved at Southampton. However, given the speed and reliability of the high-tech equipment and processes envisaged, Bamford now calculated that the new factory would be left with spare capacity of around 30 per cent. So, a revised proposal was drawn up which assumed a greater use of capacity – with no increase in fixed costs or labour – and a modest increase in market share.

In September, a revised paper was presented to the board which now offered a simple contrast between two alternative options: the 'recommended proposal' was now costed at £12.5 million, as Stokes had agreed in the interim to put in £1 million from the sector R & D budget. This was further reduced to just over £11 million by including a number of additional 'costs and returns' (including a sale and leaseback arrangement with a government agency for the Aberdare building and site, see Chapter 3). As a result the projected DCF return rose from 21 to 24 per cent. Between June and September an assumed increase in market share had emerged as a significant new element in the investment justification. However, the assumption was not based on any detailed market analysis or a revised commercial strategy, but simply on the increased capacity of the automated factory and the need to increase the DCF return. The Southampton option now involved a more substantial and immediate refurbishment of the existing factory than had been envisaged in June. As a result the DCF return went down from 48 to 28 per cent. Like the recommended option, this proposal also aimed for a good rate of return on investment, but this was to be achieved by spending as little as possible on new machines, cutting back to a minimum and maintaining the current output and market share.

The first option was clearly a higher risk strategy than the second. Many of the proposed machines and processes were untried and untested and profitability would be low in the short term. The proposal was also predicated on the need for an increase in the company's share of a market currently dominated by four big companies, with all the dangers of price wars and cut-throat competition that this implied. The board had a clear choice with two very different strategic approaches to the definition of business objectives. The first was more short-term and lower risk, with a shorter payback period and limited scope for increasing output and improving market share. The second was longer-term, with a longer payback period and higher risk, with a greater scope for increasing output but the corresponding need to increase market share in an oligopolistic market. The DCF returns over a 10-year period were now broadly similar.

In reality, of course, the decision had already been made at the informal meeting in March and at the board meeting in June, which had given approval in principle for the CIM proposal. The September board meeting approved the 'recommended' proposal unanimously.

Reflecting six years later on the board's deliberations, Arthur Stokes commented:

> The task of the finance director in this sort of set-up is clearly to present the financial aspects of any proposal in the coldest possible light so that everybody knows exactly what the situation is. There is no doubt that the board of PG, the

management of PG, the management of the cable sector, all knew perfectly well that if this project were to be carried out, it would cost more than a conventional project and the return on investment would be lower than that of a conventional project in the short run because it should have a long life. It would have substantial opportunities of improving its return as the years went by, because it was a much more efficient factory, able to respond better to the developing market need, and therefore assure PG a stronger place than it would otherwise have had in the industry. And *that* you can't ever put down into clear financial evaluation, that sort of thinking cannot be evaluated, you have to have a certain element of judgement. If one relied entirely upon discounted cash flow returns on capital investment, you wouldn't need managers, you could have a computer that did all your thinking for you!

I think the role of the board at that time was to look at all the facts, including the cold facts presented by the finance director, and to say, yes, we're taking account of all these facts, bearing in mind the needs of the sector, the development of the market-place and the risks that we may be wrong, it's worth doing, that really was the decision taken in 1985, with eyes open, risks assessed as well as they could be, we decided to go ahead with the project. (Interview, 28 August 1991)

George Sherfield's recollections had a different emphasis, but a similar conclusion:

If you slavishly follow net present value, discounted cash flow, return on capital employed, and you read all the textbooks and you slavishly follow the method of doing the arithmetic and produce reports, at the end of the day you'll go bust, because no mathematical system will take the decision for you . . . I'm not the greatest fan of DCF, it is a tool, and overriding that you have to take certain commercial judgements. But it's a very useful tool in making comparisons on profits and in making the final judgement . . . To become the lowest cost producer in a low technology product with an expensive all singing all dancing factory, there's only one way you're going to do it, you carry low stock, get your raw materials just-in-time. To pay the return on the investment, the interest on the investment, you've got to get that volume out of the door. [The September board meeting decided that] profit growth from the Aberdare CIM building wires factory could only come from volume. (Interview, 5 March 1991)

Pirelli General's experience in securing their finance director's support for an experimental technical innovation compares strikingly with that of the Japanese multinational, Sony. Initially, Sony's accountant refused to agree to a proposal to invest in Japan's first tape recorder just after the Second World War. In the end, the company's two leading executives decided to gang up on him, invited him to dinner in a black market restaurant and plied him long into the night with food and alcohol before he relented (see Morita 1994: 57–8). According to another source (*Guardian*, 22 April 1989), Sony's later decision to manufacture digital recording equipment – now one of its highest money-earners – was also apparently unjustifiable at the time by any accounting system in the world. The article concluded: 'Would British . . . engineers have enough clout, with or without a bottle of whisky, to override the money men?' At Pirelli General in 1985 they did. Whether this would still hold true in the mid- to late-1990s is another question (see Chapter 11).

## Summary

In this chapter we have seen how a small project team, composed of two PG managers and two sector technical specialists, was set up in early 1985 to investigate recent developments in computer-integrated manufacturing and to prepare the formal case for investing in a new experimental factory. At the same time, it was clear that the substantial investment required for the project would need to be costed and justified financially before it could be presented to the Pirelli General board of directors. The detailed work on the financial justification was carried out by accountant Paul Bamford, who was seconded full time to the project from November 1984 to September 1985.

In March 1985, a meeting was held in Southampton at which David Regan presented a technical update and initial costing of the project to senior sector and PG managers, four of whom were members of the Pirelli General board. There was strong support for the project as presented, although the projected cost and payback period were felt to be too high to be accepted by the board, which had three non-executive directors with industrial and merchant banking backgrounds. Regan, Bamford and the new PG managing director, Giuseppe Mancini, were charged with preparing revised proposals for the regular three-monthly board meeting in June which would compare the various options, using the discounted cash flow (DCF) technique to evaluate the return on capital invested.

At the March update meeting the Pirelli General accounts manager, George Sherfield, became concerned that the project was being driven too strongly by the engineering concerns of sector managers. He therefore argued that a cheaper and possibly more profitable option – the refurbishment of the existing Southampton factory – should be costed more fully and presented to the board in June as a real alternative to the high-tech option. This was agreed.

At the June board meeting, five options were put to the Pirelli General board of directors for approval. These included a 'recommended' option favoured by the majority involving substantial long-term investment in a new high-tech factory, and a much cheaper 'refurbished' Southampton option favoured by the finance director and showing a DCF return of 48 per cent over 10 years compared with 21 per cent for the recommended option. After intense discussion, the board accepted the recommended option in principle, but no release of funds was agreed until revised calculations had been carried out to include an increase both in the utilization of plant and market share. By the time of the next board meeting in September, the revised 'recommended' option was approved, with a smaller initial investment and more ambitious production and commercial goals.

# Questions for Discussion

1 Which of the five options presented to the board in June 1985 would you have chosen, and why?

2 Discuss and evaluate the appropriateness of the criteria used by the board in making its eventual investment decision?

3 How far do the theories and models outlined in the introduction help us understand the process of decision-making at PG culminating in the board decision of September 1985?

4 To what extent would you describe the strategy adopted as a technology-driven strategy (see also Chapter 13)?

5 Assess the role of the finance function in shaping the process and outcome of strategic innovation at Pirelli in 1984–5.

# 3
# Siting an Experimental Plant

In the previous two chapters we have looked at the literature on strategy development and on financial justifications for large-scale investment decisions. We also examined the concept of technology-driven strategies, the idea that technological innovation is the key to competitive advantage for advanced industrial nations faced with growing global competition from low labour cost countries. In the main body of Chapter 2 we saw how one particular multinational company adopted a technology-driven strategy on an experimental basis in a division of one of its national operating companies.

During the period in which this strategy was adopted, the questions of site location, building construction and plant layout were treated as important, but essentially secondary matters. In this chapter we will see how they came, gradually, to achieve a greater prominence in strategy development. The eventual adoption of a fully-fledged 'greenfield' site location became not only a crucial precondition for the full implementation of the technology-driven strategy, but encouraged and facilitated wider innovations in organization culture and human resource management.

## Thematic Focus: Greenfield Sites and the New Plant Revolution?

In 1879 the chocolate manufacturers Richard and George Cadbury were among the first of the enlightened entrepreneurs of Victorian England to move out of their cramped inner city premises to a leafy and pleasant rural setting (see Campbell-Bradley, 1987; Smith et al., 1990: 50–7). They erected a new purpose-built factory in Bournville, a small village outside Birmingham, where they also hoped to establish a 'model industrial community' imbued with the non-conformist Quaker values of hard work, mutual cooperation and abstinence from alcohol. In the early 1980s, the same Cadbury organization relocated to a greenfield site at Chirk near the town of Wrexham in North Wales. This time, according to a study of the relocation (Whitaker, 1986: 661), the primary motivations were not a mixture of commercial logic and religiously inspired philanthropy, but economic rationality and business calculation pure and simple, that is, 'the need for good communications . . ., the availability of a suitable site, eligibility for government grants, and . . . the availability of local labour willing to work a continuous shift working pattern'.

The recent academic discussion of greenfield sites can be traced back to a number of publications by the American social scientist, E.E. Lawler (1978, 1982, 1986). He was particularly interested in new trends in employee involvement and what he called 'the new plant revolution'. He argued that newly established plants in the USA had tended to introduce a range of mutually reinforcing arrangements in which 'almost no aspect of the organization had been left unturned' (1978: 6–7). According to Lawler:

> new organizations simply have a number of advantages when it comes to creating high involvement systems. They can start with a congruent total system; they can select people who are compatible; no one has a vested interest in the status quo; it is possible to do the whole organization at once. (1982: 307)

This emphasis on organizations being able to start with a 'congruent total system', a system where all parts of the organization fit and are mutually supportive, is crucial. It is one of the defining structural features of greenfield sites and one which gives them their particular potential for innovation, not just in technology, but in other aspects of business management too. The American social scientist W.F. Whyte specified this potential more precisely when he argued that it was easier to develop new forms of manufacturing organization on greenfield sites because '*the spatial design, technology and people involved are all starting fresh*' (1990: 337–48; emphasis added). Spatial design, technology, and people are the subjects of the following six chapters.

The UK debate about greenfield sites has been much exercised by problems of definition. Three different usages of the term can be identified (Beaumont and Townley, 1985: 1):

- a multi-plant organization building a new plant;
- the physical location of that plant outside a major manufacturing centre;
- the presence of new work practices and arrangements stemming from a new employee relations philosophy.

The first usage focuses our attention on one particular aspect of the operation of greenfield sites in multi-plant companies: their relationship with the wider organization. In Pirelli, as we saw in Chapter 2, the decision to build a new factory in the UK was based on the desire by senior managers at group level to set up an experimental plant from which it could draw lessons for the rest of its operations world-wide. In this sense, it was intended to be both a model and a catalyst for change in the rest of the organization. The position of a greenfield plant within its wider organization is not without its problems, as Lawler has noted, describing greenfield sites as 'foreign bodies inside larger organizations' (1978: 11). The changing relationship between the new experimental plant, the UK company (where we have already noted some tensions, even at an early stage) and the Pirelli group world-wide, will be a major theme running through subsequent chapters.

The term greenfield site is also used in a second sense, simply to describe the physical location of a plant outside a traditional manufacturing area, as in the case of Cadbury at Bournville. While this is the initial derivation of the term, it is rarely argued nowadays that a new plant needs to be on or near 'green fields' to qualify as a greenfield site. However, its degree of geographical isolation, or at least separate identity, from the rest of a multi-plant company can be important, especially when it is trying to break with the way an organization has traditionally been run.

More important for the UK debate (see Beaumont, 1985; Newell, 1991, 1993; Preece, 1993) has been the third usage of the term greenfield. Here, particular emphasis is placed on new working practices held together by a new philosophy of employee relations or, in current parlance, human resource management. This is the particular focus and interest of Beaumont and Townley, for whom the greenfield site

> offers the prospect of a *tabula rasa*, . . . the possibility of establishing work organization, job design, personnel and industrial relations policies afresh rather than attempting to tackle these issues on an ad hoc basis in existing plants. It provides the opportunity to experiment with the development of a coherent 'greenfield' philosophy. (1985: 189)

Newell (1991) has taken this further, developing an ideal-typical model of a greenfield site. She distinguishes it from a 'new plant' by virtue of its adoption of a 'new philosophy towards the way in which people are managed', particularly in the areas of training and career structures, remuneration, harmonization of terms of conditions of employment, and employee involvement (1991: 22). 'New' here does not necessarily mean the best or latest policies (although in practice it often does), but new to the organization concerned, and integrated into wider business strategies.

This usage, which is also adopted by Beaumont and Townley (1985) and Preece (1993), captures important features of the greenfield site phenomenon. It emphasizes the opportunity for managers to make a philosophical break with traditional ways of doing things and, in particular, to integrate human resource policies with wider business and corporate strategies rather than developing them in an *ad hoc* fashion. However, it also misses what is distinctive about greenfield sites in two main ways. First, there is an over-emphasis on employee relations, human resource, or people management. The result is an under-emphasis of the other areas in which greenfield sites offer the opportunity for innovation, such as building design, internal plant layout, technology, supplier relations, customer relations, and distribution and communication networks. These have been important features of greenfield sites in other companies (see, for example, the case studies of Nissan by Wickens, 1987, and Garrahan and Stewart, 1992; and of Toshiba by Trevor, 1988). They also played an important role in the Pirelli General experiment.

Second, at a more theoretical level, there is a tendency to over-emphasize the voluntarist strategies and choices of individual managers in achieving 'greenfield change' and to downplay the importance of the opportunities for change provided by structural conditions. For example, Newell argues that, as long as an organization has a greenfield philosophy (which she believes probably requires that 'at least one senior manager is "new"'), it can nevertheless be treated as one, even if it does not have all the structural attributes of a greenfield site. This leads her to argue that the distinction between greenfield and brownfield sites is not really useful:

> We should be concerned not with greenfield or brownfield sites, but with the idea of a greenfield philosophy or strategy which may be adopted regardless of the physical attributes of the plant and the labour source. Indeed, it is the very possibility that a greenfield philosophy can be introduced into an existing plant that should engage our attention, for it is this prospect that presents a much greater potential for the radical transformation of industrial relations than the opening of a significant number of new plants. (Newell, 1991: 32)

Newell is right to suggest that the adoption of an integrated human resource strategy (in her terms a greenfield philosophy) in brownfield site companies would do more to transform industrial relations in Britain, the USA and elsewhere than the opening of a small number of new plants (see Millward, 1994: ch. 4). She is also right to stress that innovation in work layout, technology and the management of human resources is possible on brownfield sites, perhaps more than many managers believe, and particularly so in times of economic crisis and turbulence. Indeed, there are some well-documented examples where organizations have turned down the option to move to a greenfield site in the justified belief that their existing site and workforce provide a better basis for achieving major changes in manufacturing and working practices (see Hendry, 1993). Finally, evidence from the USA suggests that the well-publicized experience of successful greenfield sites can be instrumental in influencing managers in existing brownfield sites to adopt greenfield-type practices in their own organizations (see Beaumont, 1985: 17).

Nevertheless, the distinctive systemic feature of greenfield sites, newly established plants, or new design plants, however they are termed, is that they provide a strategic structured opportunity (1) to design all aspects of plant operations afresh (2) to design all aspects of plant operations so that they are congruent with each other (3) to experiment with untried systems or practices. Both greenfield and brownfield sites are, to use the terminology of the American sociologist Robert K. Merton, a mixture of 'opportunity structures' and 'structural constraints' (see, on these terms, Blau, 1990: 145). Greenfield sites tend by their very nature in the direction of opportunity structures for innovation, experiment and new philosophies, while brownfield sites tend to be dominated by the structural constraints of what already exists. This is well brought out in the following passage from Kochan et al.:

> New work systems concepts are difficult to incorporate into existing union and non-union plants since they depart so dramatically from established patterns and assumptions. Moreover the consent and acceptance of workers, supervisors and middle managers is needed before such major change can be implemented in an

existing facility. On the other hand, management can act unilaterally in designing the initial work system in a new work site. (Kochan et al., 1986: 94)

Much of the UK literature veers between two extremes in its assessment of greenfield sites, either celebrating the radical changes achieved (see Wickens, 1987; Trevor, 1988) or expressing scepticism about the evidence and debunking the 'myths' (Newell, 1991; Garrahan and Stewart, 1992). In what follows we will find ample support for both points of view.

---

# Identifying Alternative Locations

The Pirelli General board of directors approved the investment in a new general wiring factory at its meeting in September 1985. It might be thought that the search would then have begun to identify where the plant could be located. In fact, this process had already begun in 1984. Indeed, most of the detailed investigations of different locations were carried out between November 1984 and March 1985, well before the board gave the go-ahead for the project.

## *The findings of the Roberts report*

The first internal assessment was made by Barry Roberts in the feasibility study report of June 1984 (see Chapter 1). In a four-page section entitled 'site location' he identified three main decision criteria, set out a number of alternatives, and came to a conclusion about the best site. The decision criteria were funding, labour, and communications.

In looking at funding, three types of regional government assistance were identified:

- development area grants (up to 15 per cent of expenditure on buildings and machinery).
- special development area grants (up to 22 per cent of such expenditure).
- enterprise zone allowances (which included a 10-year rate-free period and additional tax allowances).

Development and special development area status were determined according to levels of unemployment in a particular area, while enterprise zones were designated areas in which the government had decided to encourage industrial development.

In the Roberts report five regions of Britain were selected for comparison – North-West England, North-East England, East Midlands, South Midlands, and South Wales – each of which contained towns or areas with different mixes of development grant and enterprise zone status. Roberts assumed that the new factory building would need to be purpose-built – as no developer would be likely to have one available of the appropriate size –

and that it would be necessary to negotiate with a developer about leasing arrangements. Finally, he noted that the company already owned two sites suitable for development, at Bishopstoke just north of Southampton and at Aberdare in South Wales. The 75-acre Bishopstoke site already housed PG's telecommunication cable factory and had significant scope for further building. The Aberdare site had been bought by PG in 1971 and was being run down prior to transfer of special cables manufacture to a new facility at Bishopstoke.

In terms of labour, Bishopstoke and Aberdare had two immediate advantages over the other sites. Both had existing workforces – 330 at the Southampton general wiring factory, most of whom could transfer easily to Bishopstoke 10 kilometres away, and 70 at Aberdare – whose cable knowledge and skills would be an asset in the start-up of a new plant. Also, both options would save on the redundancy costs which would be incurred if a different site altogether was chosen (these costs were calculated at around £5400 per employee at 1984 prices).

Apart from these factors, Roberts concluded that, with the high levels of unemployment in 1984 (around 13 per cent in the UK as a whole, over 20 per cent in most development areas), availability of labour was unlikely to be a problem. Poor industrial relations were mentioned as a potential drawback in some cases, but Roberts argued that these were associated with the concentration of strike-prone industries in these areas rather than anything specific to the region or its workforce. He therefore suggested that it was best to look at the performance of their main competitor companies – BICC, Delta and AEI – in the various regions selected, and in so doing he found that their industrial relations were no better or worse than at PG's plants in Southampton and Bishopstoke. He therefore discounted industrial relations as a significant factor in the site location decision.

The report then went on to examine the cost of labour in the various locations. Here the comparison was extremely crude. Roberts compared the average labour costs of the PG factories at Bishopstoke in the affluent south of England with those at the Aberdare factory in the depressed Welsh valleys, finding that the Aberdare costs were around 16 per cent lower. He then applied the Aberdare figure to all the other regions except for the South Midlands, whose labour costs he estimated to lie midway between Bishopstoke and Aberdare.

Finally, the report looked at the differential cost of distributing the finished cable from the factory to the 10 regional depots from which they were despatched to the local branches of the electrical wholesalers (see Figure 3.1 below). The South of England, where the company's other factories and main distribution depot were situated, and the East Midlands, which was located in the centre of England and also close to the M1 motorway, were seen as the best distribution locations. The worst locations – geographically and in terms of the motorway network – were the North-East and South Wales.

The Roberts report concluded by bringing together all the quantifiable

**Figure 3.1** *Distribution depots and potential sites for new building wires factory, 1984–5*

factors – grants, site values, distribution, labour and redundancy costs – in a single figure. Aberdare was found to be the most beneficial by a small margin from the North-West and the North-East, followed by the South Midlands, Bishopstoke, and the East Midlands in that order. The first paragraph of the executive summary of the 1984 report on the new building wires unit announced confidently: 'the preferred location for this unit is at Aberdare'.

## Initial investigations of site location

When accountant Paul Bamford was seconded to work on Project X in the autumn of 1984 on his return from Canada, one of his first pieces of reading was the Roberts report. Hidden on page 26 he found a note to the effect that the government was about to change some aspects of its regional industrial policy. He made some inquiries and found that the maximum amount of regional development grant which could be awarded for each job created was about to be reduced from £12,500 to £10,000. In late October 1984 he contacted David Regan – who was not due to start work at PG until 1 January 1985 – and they decided it was worth submitting a grant application under the old regulations just in case they should eventually decide to develop the existing Aberdare site for the new factory. The application was handed in on 26 November, just three days before the changes came into force.

We will look later at the details of the application. But in drawing up the grant application, Bamford began to become more interested in the whole question of site location. Realizing that some financial assessment of the various options would have to be made before any proposal was put to the board, he decided to pursue the question further. With Regan's support, he spent much of January and February 1985 – often accompanied by Dick Brown, PG's chief engineer and a machinery and buildings specialist – travelling the country visiting areas in which the new factory might be located. Because the company had made a formal application for regional assistance to develop the Aberdare site, and was already in correspondence with local authorities over various details, Bamford decided to spend much of his initial time pursuing the Aberdare option. However, neither he nor Regan were particularly committed to locating the new plant at Aberdare, and he was given a free hand to assess the various alternatives.

Regan, meanwhile, had his own hands full trying to get to grips with the technical side of computer-integrated manufacturing systems. He knew that he would have to make a convincing presentation to Stokes and Mancini some time in March. Given Pirelli's reputation for technical know-how, and the strong engineering bias of the initial project team, he made a conscious decision to concentrate initially on developing the technical side of the project.

During January and February 1985 Bamford visited six sites in all (see Figure 3.1): Aberdare in South Wales, Corby in the East Midlands, Dudley and Telford in the West Midlands, Milton Keynes in the South Midlands, and Flint on Deeside, North Wales. In each case, he met local officials and gave an initial outline of the company requirements in terms of communications, size of workforce (estimated at around 175, just over half the number employed at Southampton) and size of site (the assumption was that it would be around 21,000 square metres, roughly equivalent to the Southampton general wiring factory). He then asked what each locality had to offer on each of these issues, as well as education facilities, grants, subsidies, and sale and leaseback arrangements for the building and site. Following each meeting he would draw up a list of pros and cons of each location. With the exception of Aberdare and Bishopstoke, location meant at this point a town or 'travel to work area' associated with a town, not a particular site. This would come later.

In South Wales, as we shall see, he met a wide range of civil servants, local government officers and industrial relations specialists. In those cases where there was an enterprise zone (e.g. Corby, Milton Keynes, Telford and Flint), he met the managers of the enterprise authority. In all cases he met senior executives from local government, who helped him with planning regulations and also gave him details of various education and communication facilities and other incentives which might be made available in order to attract investment and jobs.

In mid-February, Bamford spent a number of days back in Southampton preparing detailed calculations. By the end of February he had narrowed the options down to five. Three sites were already owned by Pirelli – Southampton, Bishopstoke and Aberdare. Southampton was not considered a realistic possibility at the time (see Chapter 1), but it was mentioned simply because it was where the existing factory was located. The other two sites were Telford and Flint. Telford represented financially the most acceptable option in terms of overall business location and distribution costs, while Flint was the best option in terms of overall size of grant (it was, unusually, both an enterprise zone and a special development area). He then compared all five options – assuming all would have an identical cost in new plant and machinery – on a range of cost criteria: land, building, training and start-up, reorganization, redundancy, grants receivable, wage benefit, rate benefit (for enterprise zones). In addition he noted that Aberdare and Flint, as ex-mining and steel areas, were eligible for special loans from the European Coal and Steel Community (ECSC), but he made no allowance for this.

Flint proved to be the cheapest option by nearly a million pounds. Aberdare and Telford were equal second, followed by Southampton, with Bishopstoke the most expensive at over £2.5 million more than Flint. But as with the decision to invest, the site location decision was not made on quantitative data alone.

# Choosing the Location for a New Plant: General Considerations

Decisions to locate new plants are sometimes bound up with decisions to invest in particular countries or trading regions. For example, recent Japanese investment in the USA and in European Union countries has been explained by their desire to gain access to what are still seen as protected if lucrative markets (see Thomsen, 1992; *Economist*, 1985). Indeed, market access is one of the main determinants in any investment decision, new site or not. However, the fact that by the early 1990s the UK had more than 30 per cent of all Japanese investment in the European Community has generally been put down to three main factors: the UK's tax regime – with one of the lowest levels of corporation tax (35 per cent) in the EC and the lowest top rate of income tax (40 per cent); its language – English is the language of business and the first and only foreign language for many Japanese business people; and strong governmental support for inward investment (see Cinnamon, 1989; Popham, 1991).

In surveys and single-company accounts of new plant locations, two additional factors are nearly always present in one form or another: resource endowments and labour. There is some dispute among economists on the effectiveness of state support for commercial development (see Wren, 1989; *Economist*, 1991a). However, all the evidence suggests that grants, subsidies and other forms of support, such as education and training provision and assistance with planning permission, play a significant role in particular investment decisions (for examples of particular companies, see Whitaker, 1986; Thomas, 1991; Garrahan and Stewart, 1992: ch. 2; for examples of particular towns or regions, see Lewis, 1988; Wheatley, 1991; Kennedy, 1991).

Attracting one major investor to a particular region can also have a bandwagon effect, as was the case with Sony in South Wales and Nissan in the North-East of England (see Hague, 1989, on 'Japanising Geordie-Land'). However, what attracted these companies initially was not only state support, but the availability of a suitable infrastructure and communications network – roads, housing and telecommunications – and appropriate land and buildings. Nissan UK, for example, required a site that was not only big enough to house its new factory, but also to accommodate its component suppliers.

The importance of labour as a factor in site location decisions is more complex and contentious. The American economist Robert Reich has argued that the existence of *skilled* labour has been one of the most important elements in US location decisions. However, Garrahan and Stewart have suggested that it was the plentiful supply of *semi-skilled* labour which attracted Nissan to Sunderland (see Thomsen, 1992; Garrahan and Stewart, 1992). Hill and Munday (1991), in contrast, have argued that the most important determinant of new plant investment in Wales has been the general earnings ratio, i.e. the relative cheapness of Welsh labour. All these examples show the importance in investment decisions of external labour market factors, such as the availability, cost and appropriate mix of skilled and semi-skilled labour.

Interestingly it appears that, when it chose its site in Sunderland, Nissan also placed great store on cooperative and flexible employee relations. Indeed, the conclusion of a single-union agreement with flexible work practices was one of the preconditions for its investment (see Garrahan and Stewart, 1992). The same was true of Toshiba's investment in Plymouth in the South-West of England (see Trevor, 1988). Indeed, when Rank and Toshiba ended their joint venture company in 1980, Toshiba's decision to stay in the UK was based not only on its assessment of the future market for televisions in Europe, but also on the belief that the company would be able to implement new working practices and a 'new' industrial relations. To achieve this, it decided to close down all the old joint-venture plants and make all existing employees redundant, thus creating 'greenfield conditions' for when the new plant opened (Trevor, 1988: 32). It then took advantage of these conditions, and the high levels of unemployment in the Plymouth area, to negotiate one of the first single-union 'no strike' agreements in the UK (see Trevor, 1988: ch. 2).

# The Choice of Location by Pirelli General

## Preparing an application for a regional development grant

Paul Bamford carried out most of his investigations into various sites for the new experimental plant between November 1984 and March 1985. His involvement with this part of the project began with the drawing up of the application for a regional development grant to site the plant at Aberdare. The request was for a (tax-free) grant of nearly £2.7 million, i.e. around 15 per cent of the eligible capital costs of the new plant, which at the time were projected to be around £16.5 million. Selective financial assistance (subject to corporation tax of 35 per cent) was also requested under section 7 of the Industrial Development Act 1982. This was a discretionary fund to encourage projects which could be expected to lead to a 'significant improvement in performance and strengthen the regional and national economy by improving efficiency or introducing new technology or products' and thus helping to provide 'more productive and more secure jobs'. The application was made on the assumption that the new plant would create 175 permanent full-time jobs by the time it was fully up and running.

In all regions of Britain apart from Scotland and Wales, such applications would have been made to the Department of Industry in London. However, both Scotland and Wales are regions with strong national identities and a heavy dependence on traditional industries such as mining, steel, and, in Scotland, shipbuilding. Against this background both had been granted their own regional 'Offices' in 1964 with a government minister of Cabinet rank. PG's grant application was therefore sent to the Welsh Office based in Cardiff, the capital city of Wales. From this point on it is possible to detect a concerted 'mobilization of bias' from within the principality of Wales which generated a powerful momentum in favour of the location of the new plant in Aberdare.

## Links with government and semi-governmental agencies

On 27 January 1985 a senior member of the industry department of the Welsh Office travelled to Southampton to talk through the grant application informally with Bamford. He followed this up with a formal letter asking for documentary evidence that, if no grant were to be awarded, Pirelli's Aberdare factory would close. Bamford replied by return, enclosing three documents. First there was a company statement to all employees at Aberdare, dated 23 September 1984, announcing the relocation of 'special cable' manufacture from Aberdare to South Hampshire, leading to the run down, and in the long term almost certainly the complete closure, of the Aberdare plant. Second, there was a letter dated 30 January 1985 from PG accounts manager George Sherfield, confirming the construction of a new special cable factory at Bishopstoke and the eventual closure of the Aberdare factory. Finally there was an excerpt from PG's 1984 strategic plan which included the following passage:

> PG has been slow to make progress [in the special cables industry] in the last few years. In addition to the commercial and technical barriers characteristic to this business, the company has mistakenly attempted to convert an old Welsh power cable works for the purpose. It is judged that to move forward with reasonable chance of success, PG must first take a step backward and move away from Aberdare. *Aberdare can be described as the wrong buildings with the wrong equipment improperly laid out in the wrong location with the wrong type of workforce.* (emphasis added)

These three documents demonstrated to the Welsh Office the real threat of the complete closure of PG's existing Aberdare plant.

As soon as Bamford had made contact with officials in the Welsh Office, they had informed their counterparts in the Welsh Development Agency (WDA) about Pirelli's application. The result was an immediate invitation to Bamford to visit the WDA Chief Executive.

The WDA had been established in 1976 with four main objectives: to further the economic development of the region; to promote, maintain or safeguard employment; to promote industrial efficiency and competitiveness; and to further the improvement of the environment. In some ways it was like an industrial development bank. However, in its early years at least, it was heavily involved in trying to generate employment. It pursued what was called at the time a 'mousetrap strategy, . . . putting down a bit of cheese and welcoming any investor – high-tech, low-tech or no-tech – that [was] attracted by it' (*Economist*, 2 February 1985: 8). By 1984–5 the WDA's budget was around £76 million per annum, and it had acquired substantial autonomy in allocating its resources. Its development arm, Wales Invest (WINvest), had made a special priority of reclaiming industrial waste land for industrial development. It also bought, renovated and erected buildings for occupation by new investors, not only from overseas, but from the rest of Britain and from within Wales itself. From January 1985 WINvest provided a kind of 'one-stop' shop for Bamford, putting him in touch with all the agencies and individuals he wanted to contact in preparing the data on the Aberdare option.

In fact, Pirelli had already approached the WDA in the summer of 1984 soon after it had decided to transfer its special cables operation to Hampshire. It wanted to find out whether the WDA would buy some or all of its land and help refurbish or develop the building for future use, either by Pirelli or by another company. As a result of this approach, the WDA had advertised in July 1984 for tenders to carry out a site investigation. The original drawings of the factory building showed that part of it had been erected above a disused coal mine and might be liable to subsidence. In late November, following Pirelli's grant application to the Welsh Office, the WDA appointed a firm of consulting civil engineers to select a contractor from the various tenders. This was done in early January 1985, and on 31 January Bamford visited the main offices of the consulting engineers in Cardiff to finalize arrangements for investigations to begin. This involved making a series of boreholes deep in the ground at various points on the site. Work began on 5 February and lasted nearly three weeks.

On 31 January 1985, the same day as he visited the consulting engineers, Bamford made the first of many visits to the head office of the WDA in Cardiff, where he met its chief executive. Events now moved with great speed. Bamford was invited back to South Wales in mid-February to meet members of WINvest. They encouraged him to return to Wales again as soon as possible to meet some more people who might be important in the setting up of the new plant. The visit was arranged for 27 February.

### Industrial relations and secret trade union consultations

In the meantime, Bamford's main contact in WINvest sent him a copy of a 20-page survey on the Welsh economy by the influential weekly magazine, the *Economist*. This had been published on 2 February 1985 and gave an extremely positive picture of investment opportunities in Wales. It challenged Wales's City of London image as 'a place where fierce unionized workers live in small terraced houses alongside derelict steelworks and slag-heaps'. Data were provided which showed the substantial recent inward investment into Wales, including a number of leading Japanese companies such as Sony, Aiwa, Matsushita-Panasonic and Hitachi. (By 1990 Wales had only 0.8 per cent of the total population of the European Community, but 8 per cent of its total foreign investment, 5 per cent of. the total UK population, but 22 per cent of UK foreign investment projects and jobs.) Above all, the *Economist* praised the Welsh labour force, which it described as loyal, adaptable and hardworking, 'egalitarian rather than Marxist', imbued with family and community spirit and used to working for one company for life. In sum, they were an excellent workforce if an employer was prepared to treat them well and to win their trust and confidence. The *Economist* also pointed out that 'greenfield' inward investors, particularly the Japanese, had always succeeded in negotiating single-union agreements with the positive assistance of the Wales Trade Union Council. This last point was underlined by Bamford in his copy of the *Economist* survey. It had clearly set him thinking.

It was thus not a complete surprise that the first meeting arranged by the WDA for Bamford on 27 February was with Peter Edwards, the General Secretary of the Wales TUC. At the time Wales was the only part of Britain where potential investors were invited to meet an official representative of the trade unions as part of an exercise to encourage them to invest. This says something about Wales, but also about the characteristics of the Wales TUC, which was described in 1980 by the then Conservative Secretary of State for Wales, Nicholas Edwards, as 'a respected part of the Welsh establishment' (HC, Oral Evidence to the House of Commons Committee on Welsh Affairs, 1980: 731–I).

The Wales TUC had been founded in 1974 on a groundswell of concern to try and do something constructive about the apparently terminal decline of the Welsh economy. While unions in the rest of Britain were rarely able to unite around positive and realistic industrial policies, in Wales unions of all ideological persuasions agreed to act together, not only to try and preserve trade unionism, but also with the aim of joining with employers, local authorities, civil servants and government ministers to promote inward investment.

For Edwards, who joined the Wales TUC as a Research Officer in 1978 and became its General Secretary in 1982, the meeting with Bamford was a cloak and dagger affair. He had been rung up the week before by an acquaintance in the WDA (with whom he had recently been on a Welsh Office trip to Japan to 'drum up business for Wales') and asked if he could clear his diary to meet someone 'about a location issue'. He had no idea whom he was meeting at 9.30 a.m. on the morning of 27 February 1985. Bamford arrived and immediately pointed out that what he was going to say was highly confidential and that if anything was made public it could jeopardize the whole project. Gradually it emerged that Pirelli General was considering investing in Aberdare, and wanted a single-union deal. (At that time, this had not been discussed at all within the company, not even with the personnel department, but Bamford had been made aware of the issue by the *Economist*'s survey and saw it as his job to explore all the possibilities.) He concluded by saying that Pirelli was considering refurbishing the old factory and hoping to run the old factory into the new. He asked for Edwards's views.

Edwards responded that the companies that had achieved a single-union agreement in Wales until then had either started afresh on a greenfield site or made a clear break between the closure of an old multi-union plant and the opening of a new one. Under such circumstances, and as long as the company allowed unions with a legitimate interest to compete openly and fairly for recognition, the Wales TUC could guarantee that all the Welsh unions would accept a single-union agreement for the new site. He then cited a number of examples of inward investing companies who had concluded such agreements and had since enjoyed cooperative and 'strike-free' industrial relations.

Bamford was impressed by Edwards's authority and professionalism. The Roberts report had discounted industrial relations as a significant factor in the choice of site location. However, Bamford was clearly worried about the industrial relations climate in one of the other potential locations, Flint in North Wales, not far from Liverpool, a city with a 'militant' reputation. In addition, in the past management–union relations at the existing Pirelli site in Aberdare had been highly conflictual, and there was still a strong collective memory within the company of a sit-in in the 1970s during which the manual workforce had locked the management out of the plant for three weeks! What had now emerged – if the new factory were to be located in South Wales and Edwards's conditions fulfilled – was the promise of a new-style single-union agreement and a new climate of industrial relations. From this time on, industrial relations became a much more central issue in the location decision.

In the experience of the WDA, Pirelli was by no means unique in attaching importance to industrial relations in deciding where to set up a new greenfield site. In 1984, the agency was asked to comment on a government White Paper on regional incentives for industrial development. This placed particular emphasis on the importance of 'wage flexibility'. The WDA responded that, in its view, this was 'not a deciding factor for incoming enterprise. It is our experience that industrial relations are a more important issue' (HC, Evidence of the WDA to the House of Commons Committee on Welsh Affairs, 1983–4: 77).

## The mobilization of local government support

Bamford's other meetings on 27 February only served to confirm the advantages of Aberdare as a location for the new plant. Immediately following his meeting with Edwards, he visited the headquarters of Mid-Glamorgan County Council in Cardiff. There he met with five senior members of the two local government bodies with responsibility for Aberdare: the chief executive, treasurer, and director of education for Mid-Glamorgan County Council, and the chief executive and treasurer of the Cynon Valley Borough Council. Bamford was impressed by the cooperation and high-level commitment of the local government leaders, and came away with a number of suggestions about the help they could offer Pirelli if it decided to locate in Aberdare. He was also reminded that two large electronics companies, Hitachi and AB Electronics, had already successfully located in the Cynon valley, and that communications between the M4 motorway and the Aberdare site were now significantly improved by the building of a new dual carriageway road from the M4 to Merthyr Tydfil and by the completion of the Aberdare bypass.

After a buffet lunch at the council offices, Bamford was driven 18 km out of town to meet staff at the Polytechnic of Wales at Pontypridd (midway between Cardiff and Aberdare). He then returned to Cardiff for a final meeting with the consulting engineers who had organized the recent site investigations at the Aberdare factory. Some subsidence had been identified, but the chances of serious further subsidence were felt to be small. They were not likely to inhibit the development of the site.

Following his return to Southampton, Bamford received follow-up letters from three of the local authority representatives he had met. Indeed, the Mid-Glamorgan director of education had prepared a specially typed briefing note (the local council had no word processors in 1985) outlining the range of educational and training establishments that would be available to the company if it wanted to locate a high-tech plant in Aberdare: the Polytechnic of Wales in Pontypridd (degrees and diplomas in electronics, mathematics, computing, business studies and management), Aberdare College (full-time and part-time courses at technician level in close collaboration with local companies), and the Pontypridd Technical College, which had particular strengths in electronics training. He also enclosed details of financial assistance offered by the county council for local collaborative projects between companies and education and training authorities. Civil servants from WINvest also offered to arrange for Bamford to meet senior managers from companies which had already engaged successfully in inward investment, and to set up a meeting between the PG managing director Giuseppe Mancini, the director of the WDA, and the chief executives of the two local authorities. The chief executive of Mid-Glamorgan County Council even offered to travel to Southampton to make a presentation to the PG board of directors.

## The selection of Aberdare

On 4 March Bamford wrote an enthusiastic memo to Mancini about Aberdare, stressing the importance of 'labour relations, employment and skills' as well as educational provision. He also suggested that Mancini should take up the WDA's offer and go and see things for himself. However, Mancini was understandably reluctant to do this until the project had been approved by the board. On 5 March, Bamford was told informally that the grant application to the Welsh Office had been approved 'in principle' to the tune of around £2.5 million. He was also informed, for reasons to be discussed below, that this sum might be increased through further negotiation. At the same time, the WDA was actively exploring the possibility of a sale and leaseback arrangement for the Aberdare site.

The first major meeting to review Project X since David Regan joined

Pirelli took place on 14 March 1985 (see Chapter 2). It was attended by senior executives from Milan and PG. A discussion of site location was scheduled for 1.30 p.m. At 11.40 a.m. on 14 March, Bamford was handed a telex confirming the WDA's willingness 'in principle' to negotiate a sale and leaseback arrangement for the Aberdare site! The WDA would buy the entire premises and provide extensions and new works, while the company would be expected to carry out repairs and general refurbishment of the building and any quality improvements that were above the WDA 'norm'. The WDA would lease back the land and premises to Pirelli over 25 years at a highly competitive rate.

At 1.30 p.m. on 14 March the 'update' meeting discussed the question of site location. The participants were presented with a map of Britain identifying the geographical location of the five main options: Southampton, Bishopstoke, Aberdare, Telford and Flint. They were also presented with an overhead outlining the detailed calculations made by Bamford comparing the five sites on a number of quantifiable measures (see previous section). He also gave a short introduction to the main non-quantifiable factors, such as industrial relations, quality and experience of labour, and education and training facilities. After a short discussion Aberdare emerged as the clear favourite. The meeting then moved on quickly to discuss the financial aspects of the project. From this time there was no further discussion of Flint, Telford or Bishopstoke.

Why was Aberdare chosen? Clearly the grant position was not as good as Flint. Yet Flint was the further away from the main geographical base of the company, and there were still some doubts about industrial relations there. Telford was undoubtedly the best business location, both as a distribution and communication centre and as a forward-looking new town with a 'green' environment and extensive new plant available for inward investors. Aberdare had a good and assured grant position, which had actually proved to be much better than on Bamford's initial calculations. Communications were relatively poor. However, there was a good supply of employees with cable experience, together with low wage costs and the guarantee of a single-union agreement. There was also a somewhat sentimental feeling among at least two senior members of Pirelli General present that, as a good employer, the company owed some kind of moral obligation to the existing workforce and to the town. Finally, everyone was impressed by the highly professional, committed and united support for the project promised by the Welsh Office, the county council, the borough council, further and higher educational establishments, surveyors, the planning authorities, and the Wales TUC, all orchestrated by the WDA and WINvest. If power is the ability to make a difference, then the Welsh state and community really did exercise power in this case.

Between March and June 1985, as we saw in Chapter 2, the debate around Project X concentrated largely on the cost of different levels of automation and investment at the Aberdare and Southampton sites. However, the question of site location was by no means fully resolved by this point. Indeed, it took a number of further twists over the following months.

# The Choice of Site

## *The availability of alternative options*

Just two weeks after the 'update' meeting, Bamford arranged for his contact at the WDA to visit Southampton to discuss the refurbishment and extension of the existing Aberdare site and the offer of a sale and leaseback arrangement. Also present was David Regan, together with a senior partner from a London firm of structural surveyors retained by Pirelli General. At the meeting it became clear that the WDA had available a number of other sites and buildings in the Aberdare travel-to-work-area and that they would be willing to consider an arrangement whereby PG would occupy one of these locations and sell the existing Aberdare site to the WDA for development. So, an unexpected and new set of choices opened up. The preliminary decision to opt for the existing site began to look much less final than it had appeared at the time.

With Regan's and Mancini's agreement, Bamford arranged to visit South Wales with Dick Brown, the PG chief engineer, to look at the other available sites. There were four in all. Option 1 was a good-size building of medium quality, piled high with sacks of dried milk. It was currently acting as a temporary home for part of the European Community's milk mountain. It was clear that, with 24-hour working envisaged for the new plant, its location near a residential area would almost certainly give rise to complaints from local residents. It was therefore discounted. So were options 2 and 3. Option 2 was a completely empty building with spectacular views down the Cynon valley, but the building was in a poor state of repair and one side had recently been blown away by a storm! Option 3 was a 'greenfield' location on a new industrial park near Aberdare, but it looked directly onto a black, ugly, smoke-belching phurnacite plant. Of greater interest was a purpose-built new factory, recently finished and originally intended for the Hoover company, whose main UK factory was in Merthyr Tydfil around 10 km from Aberdare in the neighbouring valley. Structurally it was ideal. However, it was twice as big as PG required.

## *The decision to opt for the existing PG site*

At the same time as these options were being explored, Bamford and Brown were also looking at PG's existing site in more detail. It had a number of advantages. It was situated close to Aberdare itself. The town had a pool of employees with cable experience. The local college was only a mile away and willing and able to collaborate in providing training. The building, which was about the right size, and the site, which was more than twice the size of the building, were already owned by the company. (However, this was less of an advantage than it might seem given the willingness of the WDA to provide inward investors with support for buildings and land.) The site was just 100 metres from the main Aberdare bypass, but separated from the road – and a nearby chicken factory – by a railway bridge. This meant that the company would have immediate access to the main road but with complete control – within flexible local planning regulations – over the site and its immediate environment. A location in an industrial park or residential area would have made the site subject to many more environmental and planning constraints.

Regan and Bamford persuaded Mancini to commission a surveyor to examine the site and the building and to estimate the cost of renovation and refurbishment. This work was completed very quickly. The cost was put at around £2.5 million, over half a million pounds less than the original estimate of a new building. Given that Bamford, Regan and Mancini were under pressure to reduce the initial capital expenditure for the project (in order to get board approval), this was clearly an attractive proposition, particularly if they could persuade the WDA to bear some of the costs.

Armed with all this information, Bamford suggested to Mancini that it was time to make a decision about the various options. Mancini waited until after the June board meeting, and then went to visit the various sites. It did not take him long to agree that the existing Aberdare site was the best option under all the circumstances. Regan had previously come to the same conclusion. However, he wasn't happy with the decision to refurbish and extend the existing building (built originally in 1938). He favoured a more radical solution. This, together with the conduct of complex negotiations about financial support for the project, was to occupy his, and above all Bamford's attention for the next three months.

## Informing the Aberdare workforce

During the summer of 1985 negotiations with the Welsh Office and the WDA about financial support for the project were at their height. Under these circumstances the company would have compromised its negotiating position if it had announced to its employees that the new plant would be located at Aberdare. However, it was very much aware that the remaining 70 employees at the Aberdare factory, plus the whole town and the valley, had got wind of the possibility that a new plant might come to the town. In addition, the new special cables factory at Bishopstoke was now almost ready for occupation, and an announcement would need to be made about the final run down of the special cables operation at Aberdare.

So, on 7 June 1985, the remaining Aberdare employees were given a written update by their divisional manager. The bad news was that, having already lost heavy cable production, their current main product range – special cables – was to be transferred to a new unit at Bishopstoke by the end of September 1985, producing a surplus of 46 employees. Only 24 would then remain until spring 1986, continuing to manufacture Pirelli's best-known special cables product, the fire resistant cable FP200. It too would then be transferred to South Hampshire. There was, though, some good news: 'Aberdare is now a stronger contender for the new project, and, in fact, since our last communication certain other possible sites have been eliminated'. The final bad news, however, was that if the project were to be located at Aberdare, there would be a gap of between 12 and 18 months when no work would be available. This was a heavy blow for a community which was already one of the most deprived areas in Britain, let alone Wales, in terms of income, housing and health (see Morris and Wilkinson, 1989, 1990, 1993).

# The Emergence of a Greenfield Opportunity

## *Further negotiations with local agencies*

Meanwhile, over the previous three months, negotiations over financial support for the project had been proceeding apace. On 27 March, as we have seen, Bamford and Regan met an official from the WDA in Southampton for informal discussions about the agency's offer to extend the existing building and enter a sale and leaseback arrangement over the site. On 29 March, after discussions with the company's tax advisers, Bamford sent the WDA a formal written response to the offer. He made a number of counter-proposals, including a request that PG should carry out all the extension work and that the length of the proposed leaseback arrangement should be doubled to 50 years. These suggestions were all within the WDA's guidelines. They were intended to allow the company to maximize their tax allowances on the building work. They would also have the effect of reducing the initial capital outlay on a project where all the returns would come in the medium and long term and all the initial costs in the first two to three years.

On 27 March, the same day as he met the WDA, Bamford also received a letter from the Welsh Office enclosing the formal offer of a government grant which confirmed the informal intelligence he had been given earlier in the month. The Regional Development Grant (RDG) offered was lower than had been originally expected, as it had been limited by the new rules which Pirelli General had consciously intended to avoid by making a grant application in November 1984. Under the previous rules RDGs were totally capital-based, with up to 15 per cent of the total cost of eligible buildings and equipment available in development areas, and up to 22 per cent in special development areas (the amount awarded to Nissan in support of its new factory on Wearside: see Garrahan and Stewart, 1992: 46). Under the new rules, the total amount of RDG was limited to a maximum of £10,000 per job created. The change in rules meant a difference to Pirelli of £800,000, around 5 per cent of the projected total capital cost at the time of the initial application.

Not surprisingly this led to an extended exchange of letters and a series of meetings with the Welsh Office in which Bamford attempted to negotiate an improvement in the overall grant offer. He was asked to provide detailed costings of some of the additional financial commitments associated with the project. These included start-up costs (e.g. initial training, high scrap rates with new machines), redundancy payments at the Southampton factory, and reorganization costs such as the need to pay two lots of direct fixed overheads and some variable costs during the overlap between the closure of the Southampton factory and the achievement of full production at Aberdare. (At the time this was projected to be three months – it eventually took three years.) Bamford also prepared a revised calculation of the overall capital costs. These were around £2 million less than projected in November 1984 due to the deferment of some of the investment following the 'update' meeting in March and the decision (as we have seen, later reversed) to refurbish and extend the old factory rather than build a new one. By the end of May 1985, Bamford was able to submit a revised formal request for funding which took into account both his negotiations with the Welsh Office and developments in the company's investment plans. All this, it should be remembered, took place prior to the PG board meeting of 12 June.

Meanwhile, the broad principles of the financial arrangements were becoming clearer. At the end of May, following a series of meetings, the WDA responded formally to PG's letter of 29 March, expressing a willingness to compromise on some matters – for example, the length of the lease – but not on others – for example, the rent payable by the company and the question of who should carry out the building work. At the beginning of June, the company was informed that the local council was prepared to consider a 'rate-free' period if the new plant was located at Aberdare. What was still left open, however, was the exact cost and specification of the building work to be done on the existing site. This turned out to be a crucial issue for the future of the project.

### Refurbishing the old building or erecting a new one?

As we saw, David Regan had been unhappy with the plan to refurbish and extend the old factory rather than to erect a brand-new building. However, he recognized that a reduction in building costs was an important element in gaining the board's agreement to the project, and until mid-June he was primarily concerned to defeat the proposals to defer the project for three years or even to defer it altogether and patch up the Southampton factory instead (see Chapter 2). After the 'in principle' decision to go ahead with the new plant in June 1985, he began to try and persuade his senior colleagues in Southampton and Milan to agree to a brand-new building. The role of sector management was clearly crucial in his campaign.

Regan proceeded on two fronts. First he set about persuading Mancini and Stokes that the existing Aberdare building was incompatible with sector's own philosophy for the new plant. There were a number of points in his favour. The Aberdare factory dated from the mid-1930s, the floor was uneven, and there were enormous load-bearing steel beams every 6 metres or so to support the roof. This would not only make it technically almost impossible to operate the automatic guided vehicles effectively, it would also reduce the scope for introducing a flexible, modular layout (see Chapter 4 for more details). On top of this, the high roof space with multiple steel girders would make it expensive and wasteful to heat and maintain.

On the positive side, however, with the help of the architect's and surveyor's reports Regan was able to show that they could raze the old building to the ground, level out the floor, and erect the largest modular building available at the time (16,000 square metres, 5000 metres less floor and heating space than the Southampton factory) at a cost which would not be much greater than refurbishing and extending the existing building. A new building would also reduce heating and maintenance costs, could be extended at the back if required, would give maximum scope for flexibility and experimentation, and would be fully in keeping with the forward-looking and innovative image sector wished to have for the new plant. After much debate, and a formal presentation and costing of the two options, Stokes and Mancini agreed in August to invest in a new factory. However, they insisted that the old office buildings should be retained and refurbished. (A structural mistake made during demolition ensured that the offices eventually had to be pulled down too!) Following the decision to erect a new factory, negotiations with the Welsh Office and the WDA were concluded speedily and an agreement reached which took into account the dating of Pirelli General's initial grant application.

## The emergence of a full greenfield option

For Regan, the issue of site location had appeared initially to be secondary to the detailed financial justification and technical elaboration of the project. The process of identifying alternative locations, choosing the old Aberdare site, and conducting detailed negotiations with various authorities in Wales, had generated substantial resource endowments for the project. In the end the Welsh Office provided grants of over £2.5 million, while the WDA agreed to pay 50 per cent of the cost of the new building (which was estimated to cost around £2.5 million) plus £200,000 for the lease of the site from the company for a period of 50 years. In return, the company agreed to pay an annual rent to the WDA which would represent a standard fixed return of around 8 per cent on the capital it had invested. The rent was fixed at £100,000 per annum for the first five years, after which it would be reviewed in the light of the changed market value of the site. Finally, Pirelli received additional financial assistance and grants from local government authorities (for a 'rate-free' period) and the national government's Manpower Services Commission (for the initial education and training of staff).

The debates and negotiations over site location had led to incremental developments in the overall strategy for the new plant, particularly in the technical and human resource fields (see Chapters 4–8). Regan now had the freedom to design and lay out the new site to meet all the operational and technical requirements of an advanced computer-integrated manufacturing system. The emergent decision to insert a gap of between 12 and 18 months between the closure of the old plant and the opening of the new one also opened the way for the development of a radically new human resource strategy. The company now had the opportunity to recruit a completely new workforce and to achieve a single-union agreement on its own terms. While the decision to opt for a greenfield site was implicit in the philosophy behind the new plant, the pressure to reduce overall project costs had threatened to dominate the question of site location and to undermine some of the most important structural opportunities for innovation. By October 1985 Regan had all the spatial, technical and human resource opportunities afforded by a full-blown greenfield site.

## Summary

In this chapter we have followed the complex processes which led to the eventual decision to site the experimental factory in a brand-new building at Aberdare in the South Wales valleys. Initial investigations, which began well before the board's decision to go ahead with the project, identified three sites already owned by Pirelli General and five other locations which offered various degrees of central and local government assistance. At this time, there was no commitment from PG to locate the new factory on a 'greenfield' site. Project accountant Paul Bamford visited various parts of the UK to assess the advantages and disadvantages of the various options, and he was particularly impressed with the professionalism, the extent of financial support, and the degree of local collaboration he found in Aberdare and in South Wales more generally.

At the 'update' meeting in March 1985, a preliminary decision was made to site the new plant at Aberdare, and to refurbish and extend the company's own existing factory building. However, from June 1985, once the decision 'in principle' had been made to invest in an experimental factory, David Regan set about persuading senior national company and sector managers that a brand-new factory building would be not much more expensive while being more compatible with the philosophy and technical requirements of the new plant. By October 1985, the argument had been won, and negotiations for financial support were concluded with various agencies. Although one of Pirelli General's existing sites was to be used, the outcome was, to all intents and purposes, a full 'greenfield' site option.

# Questions for Discussion

1  What are the main advantages and disadvantages in locating a new plant on a greenfield site?

2  How would you explain the high level of inward investment in Wales in the 1970s and 1980s?

3  Which location would you have chosen for Pirelli's new plant, and why?

4  Discuss the most important reasons for the company's decision to locate its new factory on its existing Aberdare site?

5  Assess the importance of labour market and industrial relations factors in locating a new plant.

# 4

# Choosing the New Technology

## Thematic Focus: Computer-Integrated Manufacturing and Automation

Henry Ford launched his Model T car in 1907 in the USA. By 1923 nearly 20 million had been sold. The Model T is a prime example of mass production, the manufacture of large quantities of similar products assembled from precisely machined components with high rates of production output. In its extreme form the customer has virtually no choice. As Ford used to say about the Model T: 'You can have any color, as long as it's black.'

In contrast, batch production involves the manufacture of small-, medium- or large-sized lots of different items – 75 per cent is in small batches of less than 50 (Greenwood, 1988: 2). Batch products are either manufactured only once to satisfy a specific customer, or produced at regular intervals to meet continuous or intermittent demand. In the latter case, companies build up stocks of an item, change over to other items, and then change back to produce the first item when stocks run low. By definition, an organization specializing in batch production has a capacity greater than the demand for any one product, so that the machines (and people) need to be more flexible and multi-purpose (Groover, 1987: 18–20).

Over the past 75 years, new production systems such as flow-lines, and new technologies such as large mainframe computers, have tended to be too costly or too inflexible to meet the demands of small- or medium-size batch production. However, microelectronics-based computer technologies now make it possible to bring the economies of scale of mass production, which constitutes only around 15 per cent of total manufacturing output, to the greater economies of scope of batch production, which makes up around 40 per cent (Greenwood, 1988: 1–2).

Different theories have been advanced to explain this change. The most prominent is the 'flexible specialization' (or 'flec spec') thesis, first elaborated in 1984 by Michael Piore and Charles Sabel of MIT. According to this theory, 'Fordist' mass production techniques are less appropriate to a market-place where consumers demand ever greater choice and product variation. For Piore and Sabel, the key to competitive advantage in contemporary manufacturing companies – whether engaged in mass, batch or unit production – resides in their ability to adapt quickly and efficiently to new product markets. For this they need flexible technologies and flexible workforces. For example, in the automobile industry, 'instead of producing a standard car by means of highly specialized resources – workers with narrowly defined jobs and dedicated machines – the tendency is to produce specialized goods by means of general purpose resources – broadly skilled workers using capital equipment that can make various models' (Katz and Sabel, 1985: 298).

This is not the place to evaluate in detail the 'flec spec' thesis, or the associated concepts of Fordist and post-Fordist production (on this see McLoughlin and Clark, 1994: 48–55). However, it is important to note that, underlying the whole debate, is the tension, and changing balance, between the requirements of consumption and production:

> Customers almost never issue orders at the same rate that companies want to make the product. Customers want maximum choice of options on products and frequently seek new options. Manufacturing wants stable plant loads, a constant labour force, smooth batch material flow, and standard products. In short customers want flexibility and manufacturing wants stability. (Plossl, 1987: 15)

According to Piore and Sabel, intensified international competition and growing consumer sophistication have caused the pendulum to swing away from the producer to the consumer. The trend towards organizational flexibility and consumer responsiveness also places new burdens and costs on manufacturing companies, whether in the form of additional equipment and staff to meet peak demands, larger inventories of finished goods and raw materials stocks to buffer demand fluctuations, or the need to provide price and service inducements during slack periods. Contemporary manufacturing companies have to strike a balance between changing customer demands, market share, and keeping fixed costs to a minimum. For many, the answer lies in technical innovation and the introduction of some form of advanced manufacturing technology.

Before we can appreciate the capabilities of advanced manufacturing technologies, we need to understand the core elements of the manufacturing process itself. Whatever the type of production (mass, batch, unit or process), technology (automated, mechanized or manual), or product (car, cable or cake), the physical activity of manufacturing involves the completion of certain basic tasks in order to convert raw materials into a finished product (see Groover, 1987: 20–4). These tasks are listed below, together with illustrative examples:

- *Processing*: transforming raw materials, e.g. metal into car panels, copper into electrical wire, flour, water and butter into dough.
- *Assembly*: combining separate work pieces into a whole product, e.g. separate panels to make a car, adding a plastic sheath to copper wire to make a cable, adding a filling and icing to make a cake.
- *Materials handling*: ensuring raw materials are available at each stage and moving the product from one stage to the next.
- *Inspection and test*: quality management of the product to ensure it meets the required standard.
- *Control and coordination*: controlling individual processes and assembly

operations, coordinating plant-level activities (labour, maintenance, materials handling, operating and energy costs).

In addition to these basic production tasks, there are a number of pre- and post-production activities. These include sales and marketing, order receipt and processing, product design and costing, materials planning and purchasing, production planning and sequencing, as well as storage of finished goods, dispatch and distribution, customer billing, and accounts. Finally, every manufacturing facility needs to be designed and laid out in a way which enables these various tasks and functions to be carried out effectively. The history of technical change in manufacturing since 1945 is a story of computerization and integration of an ever greater number of these activities (see Greenwood, 1988: 8–12).

'Computer-integrated manufacturing' means essentially that all of a manufacturing company's operations are automated through an integrated system of computer hardware and software (see Groover, 1987: 721–2). In this sense, full CIM implies a computer-integrated business (CIB) in which the output of one activity serves as the input to the next, forming an integrated system through a chain of events that starts with the sales order and ends in the receipt of payment for the goods delivered to the customer. Earlier applications of modern computers to manufacturing, such as CNC machine tools and CAD systems, were standalone 'islands of automation', in which computer control was substituted for manual control or operation. CIM systems, in contrast, are not 'substitution' but 'integrating' technologies, linking previously separate operations into a total system which is more than the sum of the separate automated parts (see Bessant, 1991, for further discussion).

The architecture of a CIM system is normally based on a hierarchical design (see Groover, 1987: 766–7; Greenwood, 1988: 15–18). Different terms are often used to describe the various levels in the CIM hierarchy, but a schematic representation of the four main levels and their core elements is presented below:

- *Business Management System*
  - sales order input and processing
  - dispatch/distribution
  - billing/sales ledger
  - financial accounting
  - personnel (payroll/attendance/training)
- *Manufacturing Management System*
  - product design and costing
  - bills of material
  - basic structure and routing of different products
  - material requirements planning (MRP)
  - purchasing
- *Plant Management System*
  - batching and sequencing of particular orders

- work-in-progress inventory
- control of materials handling and workflow
- quality control and management
- management accounting
- *Process Control System*
  - sequence of operations on individual machines (machine set-ups, production steps)
  - process control (CNC)
  - devices (machine tools, sensors, dies)

The main features of CIM systems, and the choices organizations have to make when introducing them, will be elaborated further when we look at the system chosen by Pirelli General for its new Aberdare factory. But some important general observations can already be made at this point. First, CIM is best described as the integration of the bottom three levels of the hierarchy, while a computer-integrated business (CIB) requires the integration of all four. Second, 'island' automation of particular elements within each of the four levels – e.g. billing, purchasing, production planning, process control – is quite conceivable without there being any integration between them. However, the more an organization wishes to integrate the various elements of its operations, the more important and expensive is the software and the greater the risk of failure if implemented in 'top down' fashion before the constituent parts are operating successfully in their own right (see Jones, 1988: 458). The most important and complex part of the system hierarchy is undoubtedly the plant management system (PMS). It provides the crucial link between machine automation on the shopfloor and the wider manufacturing and business management systems. It is also 'the largest single risk element' of a CIM innovation, accounting for between 25 and 40 per cent of the cost and 50 to 75 per cent of the total risk (Groover, 1987: 158–9).

The quest for a fully integrated CIM system raises further crucial questions about implementation. For example, who is responsible for the design of the PMS and its integration with the other elements of the structure? The core elements of the physical manufacturing process are normally designed, installed and commissioned by teams of equipment suppliers and in-house engineering specialists. They normally have expertise in machine systems and products, but little knowledge of software and the integration of manufacturing and business systems. In contrast, since few companies have sufficient in-house expertise, CIM systems software is frequently designed by external consultants. They are normally experts in systems design and computer programming, but have little appreciation of the intricacies of the manufacturing environment – product and process – for which they are designing the system. This 'polarization of expertise' (Greenwood, 1988: 159) between production and systems/software engineers is one of the prime reasons why advanced manufacturing systems have often not performed as well as originally expected (see Kearney, 1989; Weatherall, 1989).

It is important to recognize too that CIM systems cannot be bought 'off the peg'. Companies not only have to choose and assemble the appropriate system parts, but also design a systems architecture which will integrate them together. In 1985, when Pirelli began to design its own CIM system for Aberdare, Ingersoll Consultants had to give up a proposed study of CIM in UK plants when their researchers reported: 'we can't find any' (quoted in Jones, 1988: 458). Even by the early 1990s, the prevailing view of the major world-wide software vendors, such as IBM, Hewlett Packard and the Digital Equipment Corporation, was that 'only parts of CIM exist today', mainly because of 'connectivity and compatibility' problems between the different system components (Brown, 1991). As a result CIM systems tend to have a very high 'originality level' (Pelz and Munson, 1982). With any major innovation, particularly in technology, organizations have to decide whether to design their own solutions (invention), adapt features of solutions provided by an outside source (adaptation), or buy off-the-peg solutions from an outside source (borrowing). While most large-scale innovations will have elements of all three, the greater the element of invention, the greater the expense and the risk, and the greater the potential gap between the design ideal and the eventual operational reality (on this see Gerwin, 1984: 68–70).

The answer to many of these design and implementation questions depends on the objectives an organization hopes to achieve through computer integration. In the mid-1980s, in line with much of the engineering press and the hardware and software vendors, the companies that embraced CIM did so with the intention of being able to switch instantaneously between different products in order to meet customer orders just-in-time (JIT). In practice, full JIT production was either not attempted or, where it was, led to increased scrap and wastage and major production delays as a result of continual product switches (see Jones, 1988: 458). In contrast, more successful implementations of CIM, including a recent Japanese example in the UK, have not attempted to gear production totally to customer orders and to eliminate stocks completely, but have concentrated on enhancing overall control of the manufacturing process and reducing inventory to the minimum (see Kurimoto and So, 1990). In the same Japanese case, it was also found that 'human factors and flexible working practice [were] the keys to the successful implementation of advanced manufacturing technology' (1990: 10), a point to which we will return in Chapters 6 and 7.

This brings us to a final major issue in CIM design and implementation: the balance between automation and human intervention in the production process. In the early and mid-1980s, the holy grail of technical innovation was full computer control of all aspects of production and its correlate, the 'workerless factory'. For example, an article in the US magazine *Fortune* published in February 1983 described computerized manufacturing systems as the 'ultimate entrepreneurial system', enabling 'virtually unmanned round-the-clock operation' (cited in Forester, 1985: 285). As a leading member of Ingersoll Engineering Consultants recalled:

Ten years ago, the phrase Factory 2000 inspired futuristic visions – technology-driven, clinical, robotic factories, paperless offices, minimal workforces, major capital investment, capabilities as yet undeveloped. The focus was heavily on the mechanistic, with many a manufacturing technique heralded as the answer to competitive pressures in the future. (Small, 1992: 1)

This conception of CIM led to some enormous and costly mistakes. The most spectacular failure was at General Motors' giant Hamtrack plant in Detroit. Production lines ground to a halt for hours at a time while technicians tried to debug software. Robots, when they did work, fitted the wrong equipment, smashed cars, sprayed paint everywhere, and even began to dismember each other. It was only when employees took the initiative themselves to visit machine suppliers and develop new production techniques which enhanced machine reliability that things began to improve (see the *Economist*, 1991b).

Many similiar, if not so extreme, experiences of machine unreliability and over-complex software, together with regular failures to build in planned human intervention and appropriate organizational changes, dissuaded many US and UK companies from investing in high levels of computer integration in the late 1980s. Indeed, some commentators suggested at the time that top-down, highly integrated computer systems were no longer appropriate and had now been overtaken by less expensive, less experimental, more bottom-up initiatives in factory automation, often described as flexible manufacturing systems or FMS (see Lowe, 1990; Martin, 1989). On the other hand, by the late 1980s some companies were already finding ways of building human safeguards into CIM systems, opting, as GM eventually did, for a balance and partnership between computers and people rather than the subordination of one to the other (see Martin and Jackson, 1988). Recent survey research has confirmed the view that the most commercially successful companies using advanced manufacturing technology are those that link the design of new technical systems with the design of new organization and work arrangements (see Bessant, 1993). The main advantages and disadvantages of CIM systems are summarized in Table 4.1.

Computerized manufacturing systems such as CIM and FMS operate on a continuum between two related but potentially conflicting organizational principles, flexibility and integration (see Jones, 1988). Most factories operate somewhere between the two extremes. At the one extreme, which we might call full CIM, there is an emphasis on centralized computer control of the sequence and content of highly standardized production operations, with little flexibility at the level of the workstation or machine. At the other extreme, which we might call a low level FMS, there are high levels of decision-making at the point of production (workstation or cell), made possible by access to centrally coordinated but locally distributed systems of information. In terms of implementation strategies (see Chapters 5 and 7), an emphasis on full integration favours a more rigid top-down approach, whereas an emphasis on flexibility suggests a more phased incremental approach, building up to full integration only after each component in the

**Table 4.1**   *Main advantages and disadvantages of CIM systems*

| Advantages | Disadvantages |
|---|---|
| 1 Improved capital/equipment utilization | 1 Still relatively new technology |
| 2 Reduced work-in-progress and set-up | 2 Good design/implementation expertise hard to find |
| 3 Reduced throughput and lead times | 3 Complex systems require lengthy operator training |
| 4 Reduced inventory and smaller batches | 4 Systems are expensive |
| 5 Reduced staffing | 5 Systems take several years to implement |
| 6 Design changes can be readily accommodated | 6 Large amount of once-off software required for operations management control system |
| 7 Consistent quality | |
| 8 Reduced risk of product failure | 7 Difficult to integrate devices/ equipment from different manufacturers |
| 9 Accurate management control | |
| 10 Improved market image and credibility with customer | 8 Difficult to find system suppliers to provide long-term support |
| 11 Reduced floor space requirements | |

*Source*: Adaptation of Greenwood, 1988: 22

system has achieved some degree of reliability and stability. In this chapter we will trace how Pirelli project managers developed and implemented their own conception of CIM/CIB between 1984 and 1988. In Chapter 5 we will then see how they adapted their original strategy between 1988 and 1990 in the light of experience.

# Selecting New Machine Systems

## *The technical project team continues investigations*

By the end of May 1985, David Regan and Giuseppe Mancini were confident that the PG Board would give the formal go-ahead for Project X at its June meeting. Following consultation with Luciano Balandi, the influential head of the sector technical department in Milan, they decided to convene a meeting in Southampton on 17 June, three days after the board meeting, to agree a programme of future work on the project. The main invitees were the technical specialists who had formed the backbone of Arthur Stokes's industrial automation steering group since the autumn of 1984; Balandi, head of the sector technical department; the two members of Balandi's staff who were members of Regan's technical project team; the heads of PG's corporate technical and engineering departments, together with its chief engineer; and Ian Scott, the divisional manager of the Southampton general wiring factory. They were joined by the two UK members of the project team, Gordon Peters and Paul Bamford. Apart from Bamford, who was on temporary secondment from the accounts department, the meeting was composed entirely of experts in machine and process automation – there were no sales, marketing or personnel specialists, and no computer systems or software engineers.

In the event, as we saw in Chapter 2, the PG board of directors only agreed to the project 'in principle', sanctioning no expenditure of money until the proposals had been re-evaluated and re-costed. However, in his opening remarks at the meeting, Mancini argued they should now give the programme more formality, define sub-projects, and assign tasks, resources and responsibilities. In particular they should set about ensuring that the major process innovations intended for the new factory would actually work in practice.

Regan's minutes of the meeting – headed 'CIM Automation Project' – identified 14 actions. All but one involved work on different kinds of machines or process automation (the bottom level of the CIM hierarchy). The plant, manufacturing and business management systems (the other three levels of CIM) were not mentioned at all. From June to December 1985, the four-man technical project team concentrated almost exclusively on the more detailed specification of the new machines and machine processes.

### Machine tandemization and selection of machine suppliers

The main engineering principle they adopted in order to enhance machine automation was *tandemization*. The best way to understand this is to compare it with the more traditional methods of cable production. The manufacture of general wiring cable begins with a large drum of 8 mm copper wire (rod) and ends with drums or boxes of finished cable for dispatch to the customer. Production is made up of four main sequential processes:

- *Metallurgy*: forming the copper conductor (wire) into the required sizes and types.
- *Extrusion*: insulating and putting a sheath on them.
- *Assembly*: combining the components into a finished cable.
- *Rewind and packing*: cutting up the finished cable into the required lengths, and rewinding it onto the appropriate size drums, or putting it in cardboard boxes, ready for dispatch.

Producing building wires in the conventional way involves a whole number of fixed processes. In metallurgy, for example, each wire is drawn down separately and then put through an annealing process in which an electric current is passed through it so that it becomes soft and does not stretch and snap. The separate wires are then bunched together, annealed again, and then twisted (stranded) into the right shape. Traditionally, each of these processes has been carried out by a different machine, with research and development effort concentrated on improving the efficiency of separate machines and machine parts so that the wire can be drawn (or insulated and sheathed) at even faster speeds. A technical fascination with increasing machine speeds was very much part of the culture of the cable manufacturing industry world wide, and of Pirelli General, too.

By the early 1980s, however, equipment manufacturers were beginning to explore alternative ways of achieving shorter product cycles, taking the first steps to link together (tandemize) the traditionally highly separated and labour-intensive processes. When the technical project team was considering such developments, the experiences of one of the sector technical specialists and PG's chief engineer were particularly valuable. Both had first-hand knowledge of factories with continuous process 'multi-line' machines in metallurgy, which could take in up to eight wires at once, bunch them, anneal them and twist them – all in one extended machine process – delivering the completed conductor onto a large bobbin (drum) without any need for human involvement or materials handling. During 1985 and 1986, members of the project team, the heads of the sector and PG technical departments, and PG's chief engineer spent many days with leading European equipment manufacturers, encouraging them to be as experimental as they could in tandemizing previously separate machine processes.

In most areas of the new factory, tandemization involved an extension of existing machine designs rather than the design of totally new machines. However, in one crucial case – extrusion – it did. The project team, with strong encouragement from Balandi, decided to experiment with a system which was very much in its infancy in cable manufacturing, namely multiple extrusion through gear pump technology. The idea was to feed two or three formed pieces of wire conductor into the extruder and to sheath each one – and cover the sheath with a separate colour – in one gear-pump-driven machine process. Conventionally, improvements in extrusion performance involved getting the insulation lines and coloration processes to run faster, while still sheathing and colouring only one conductor at a time. The new machine was designed to run at a moderate speed, thus avoiding the quality problems and increased scrap rates often associated with faster speeds. However, because it carried out the sheathing and colouring of three conductors simultaneously, it could complete the overall task in a much reduced time. Also, if there was a sudden requirement to make a quick change from one batch (size, type, colour) to another to meet a particular customer order, this could be done by 'simply' typing the new machine set-ups into the computer, whereas in the past it would have been necessary to change the set-ups manually on a number of separate machines.

In the very early project team discussions prior to the 'update' meeting in March 1985, the largest amounts of money to be invested in new machinery were allocated to extrusion, metallurgy and assembly, in that order. Rewind and packaging had always been the Cinderella areas of cable production.

When project costs had to be scaled back between March and June 1985 in order to get board approval, the project team decided to recommend refurbishing the comparatively new packing machines in the Southampton plant rather than buying the most up-to-date technology.

After extended negotiations during 1985, the machine contracts were put out to tender by the PG corporate technical department in the winter of 1985–6. The main orders were placed by March 1986. Two machine suppliers (both German) were awarded contracts for the metallurgy area, two in extrusion (one British, one Swedish), two in assembly (one French, one British), while orders for the packing and rewind equipment were placed with one Swedish and two British companies.

## Computer process control

Machine tandemization represented a significant step in the creation of a modern high-tech manufacturing system. But if the ultimate aim was to link the various tandemized machines into an integrated process control system (the first level of the CIM hierarchy), each individual machine would need to be computer-controlled and also capable of interfacing with all the others. After consultation with the machine suppliers, the project team opted for computer control via programmable logic controllers or plcs. Plcs are mini-computer systems with a screen and a keyboard (like a word-processing system) which can be programmed – either at a distance or on the shopfloor – to direct a machine to perform in specific ways. By 1985, they were a well-tried device for computer control of individual machines, with over 60,000 in operation in a quarter of all UK factories with over 25 employees (see Northcott, 1986).

Regan and his colleagues were fortunate in being able to learn important lessons about the implementation of plc systems from visits to a number of 'reference sites' with advanced manufacturing technology (AMT). These had been recommended by the Department of Trade and Industry (DTI) in London, which held a register of companies who had successfully introduced AMT systems, often with government assistance. Virtually all the companies they visited who had wanted to advance beyond island automation of individual machines had suffered from the lack of standardization in computer controls. This had led in most cases to major difficulties in interfacing – the so-called 'electronic handshake' – between machines. Strongly influenced by this experience, the project team decided in late 1985 to standardize all the machines on one plc system, one keyboard and screen layout, and one form of plc software. The contract for the software was awarded in 1986 to a UK company with years of experience working with the plc system selected.

## Contractual arrangements with equipment suppliers

By the end of 1985, then, there had been thorough and systematic planning of machine tandemization and the process control system for the new factory (level 1 of the CIM hierarchy). What had not been resolved at this stage was who would have responsibility for the effective functioning of each machine, not only in standalone mode, but when it needed to interface with the rest of the system hardware and software. It has often been argued that the only way to achieve this is to appoint a prime contractor or 'single source coordinator', and to give them the responsibility for achieving the interfaces (see, for example, Greenwood, 1988: 243–6). However, Pirelli General, part of a multinational company infused by a culture of engineering and technical excellence, had never conceded such a role to an outside company when setting up a new factory. In this case, too, neither sector management nor the Aberdare project team ever considered such an option.

Almost by default, then, the internal coordination and integration of the machine and process control systems was left totally to the Aberdare project team and specialists from PG's and sector's technical departments. At this time, however, no one involved in project management was a specialist in the design of computer software systems. So, while the contracts agreed between PG and the machine suppliers were highly specific on such requirements as the standardization of plc hardware and software, they were unspecific and open-ended about the mechanisms for effecting the interfaces between individual machines and the rest of the CIM system. Most importantly, there was no indication as to who was to be responsible if faults on one particular machine were to bring the whole of the production system to a halt or if machines needed to be modified to achieve system integration.

### Materials handling through automatic guided vehicles

The final element of the process control system to be specified was the automation of materials handling. In fact, the choice of automatic guided vehicles (AGVs) had already been determined by sector management (Stokes and Balandi). There was only one major company in the UK which was prepared to bid to supply the AGVs, and after detailed negotiations they were awarded the contract.

AGVs are the most futuristic form of materials transport in advanced manufacturing systems. In terms of automation, they are on a par with robots and continuous conveyors. They are battery-driven vehicles which operate along predefined routes carrying quantities of product or raw materials from one location to the next. Normally they are microprocessor-controlled, and operate along a wire-guided system buried around 3 cm below the factory floor. The vehicles have individual sensors on either side which allow them to track the wire and receive instructions from a central computer (see Greenwood, 1988: 119–20).

AGVs are more expensive than traditional forms of materials handling, but more flexible in their routing possibilities. In most cable factories, and in all of Pirelli's other plants, materials transport was carried out via manually operated fork-lift trucks. If the choice of material transport had been left to the project team and Pirelli General technologists, there is no doubt that they would have opted for fork-lift trucks. These were a tried and tested technology and, although more labour intensive, much cheaper in terms of initial capital outlay. What was not possible with forklifts, of course, was computer control of materials handling and just-in-time production – the ideal of computer-integrated manufacturing to which the project team was committed.

## Designing Computerized Management Systems

When Arthur Stokes set up the industrial automation steering group in Milan in September 1984, he made sure that it included not only cable technology and machine automation experts, but also two specialists in electronics and computer systems from within the Pirelli group. One of the first initiatives taken by the steering group in January 1985 was to commission a UK software house to prepare a computer simulation of the maximum possible level of systems integration in the new factory.

By March 1985, however, it was becoming clear that Regan and the technical project team were spending most of their time on the machine and process control side of the Aberdare project. Stokes therefore decided to establish a separate 'systems' group in Milan, composed of electronics and computer specialists within Pirelli, together with representatives from IBM Italy and Elettronica San Giorgio (ELSAG), a subsidiary of the state industrial holding company IRI. Their task was to explore the wider potential of new computing and information systems for cable manufacture world-wide. Over the next few months Stokes and Balandi encouraged the UK building wires project team to meet a number of times with the new sector 'systems' group to talk through the systems requirements of the new factory.

The four main members of the Aberdare project team were all specialists in cable-making, plant installation and general management. However, none had much knowledge or experience in electronics or computer software. The same was true of Mancini and Balandi. Perhaps partly for this reason, when the project was costed prior to the presentation to the board in June 1985, the estimates for the new machine systems were put at over £8 million, while the plant management system – the second level of the CIM hierarchy which controls and integrates the various machine systems with the wider CIM system – was estimated to cost only £800,000 and the materials handling system – the AGVs – £200,000.

The main reason for setting the June 1985 estimates so low – at the update meeting in March the PMS and the AGVs had been costed at £1.6 and £1 million respectively – was to reduce the overall cost of the proposal in order to secure board approval. It also reinforced the argument that Milan should provide a subsidy from the sector research and development budget in recognition of the experimental nature of the project. In the event, Stokes provided a total of £1 million from the sector R & D budget for the project.

However, given the high cost of the AGVs, this eventually left less than £1 million for the development and installation of the plant management system, or POMS (plant operations management system), as it came to be called.

By September 1985 the IBM and ELSAG members of the sector systems group in Milan had become so convinced of the potential future market for their work that they decided to set up a joint venture company, SEIAF (Sistemi Elettronici ed Informatici per l'Automazione della Fabbrica) specializing in supplying electronic and information systems for factory automation. Stokes and Balandi supported the initiative, but decided that, as systems software was not part of Pirelli's core business (a corporate strategy issue), they did not wish to participate directly in the new company.

By this time, Regan was becoming increasingly aware of how little 'systems' input there had been into the project so far. Not being a systems or computer software expert himself, he approached Andrew Slater – head of PG's recently established corporate information systems department (ISD) – to see whether he could help. Slater knew immediately where he could secure some specialist advice and support for Regan. His deputy and systems development manager, 35-year-old David Wolstenholme, had long been fascinated by the potential of computer systems for factory automation and was ready for a new challenge. Slater introduced Wolstenholme to Regan, they hit it off immediately, and in late 1985 Wolstenholme was seconded full-time to the project. Meanwhile, Gordon Peters, who had been involved in the project from the beginning, was becoming increasingly interested in the systems side of the project, and when Wolstenholme secured two more part-time secondments from the PG corporate technical department, this made an Aberdare systems team of four.

At this time, 'systems' was a bit of a dustbin category. It described everything to do with CIM that was not concerned with machines and machine process controls – the focus of the 'technical' project team. It included everything from factory layout and systems architecture to the more detailed design and development of the higher level components of CIM: the plant, manufacturing, and business management systems (see p. 70 above). Faced with this mass of systems requirements, Wolstenholme decided that his team would concentrate on what he felt were the priority areas: plant layout, systems architecture and, above all, the plant management system. As a member of both groups, Gordon Peters provided an important link between the work of the 'systems' and 'technical' teams. However, the result of this specialization was that the detailed design and specification of the manufacturing and business management systems were left largely to Slater and his staff in the corporate information systems department.

In many ways, this was a logical division of labour given the enormous workload, the background, the jobs and the location of the individuals concerned. Machines and process control, plant layout, systems architecture and the plant management systems were technical matters over which members of the Aberdare project teams (technical and systems) had substantial autonomy. In contrast, purchasing, order intake, billing, distribution, financial accounts, and payroll, were all matters in which there were varying degrees of corporate control and where the degree of project management autonomy was inevitably more limited. The consequence was that these areas of computer systems development – particularly the commercial, financial and distribution systems – were not accorded a high degree of priority and interest by the Aberdare project teams and were certainly not subject to the same degree of project management coordination and system integration. In this sense CIM, rather than CIB, was becoming the narrower focus of the Aberdare project managers.

### Systems architecture and layout

With a greenfield site, a new building and no massive steel pillars, David Wolstenholme and his team did not need much convincing to adopt the two principles which dominate the modern factory (and CIM/FMS) literature: zonal layout and modular design. The zonal principle involves laying out the production process sequentially to bring together all the machines in each area or zone of production. By purchasing the most up-to-date equipment and tandemizing what had been until then separate machine processes, the project team was able to reduce by around a half the overall number of machines required to produce the projected volume of cable. This lean design also allowed substantial room for staff to move between machines and machine areas and facilitated the smooth operation of the automatic guided vehicles.

The actual layout chosen is best explained by describing in a little more detail the four main stages in the manufacture of a building wire: metallurgy, extrusion, assembly, and rewind/packing. At the heart of the process is a conductor of electricity, generally made of copper. The manufacturing process starts with the production and delivery of a large (5-tonne) drum of copper wire conductor. The 8 mm-wide 'rod' of copper wire has then to be reduced or 'broken down' – often in a number of stages – into the requisite size. It is then bunched or twisted together to form the basic strand. At the end of this stage, called conductor formation or *metallurgy*, the finished strand or conductor is taken up on a drum or bobbin for transport to the next stage.

The function of the next stage is to provide the conductor with insulation and sheathing through the process of *extrusion*. The way this is done is to 'pay in' the conductor through an Archimedean screw (or two screws for continuous process production), cover it – at the most simple level with a coating of plastic or polyvinylchloride (PVC) – and then 'take it up' at the end of the process onto a drum or bobbin for transportation to the next stage.

At the end of the extrusion stage, there are three possible options, depending on the type of cable. Some multicore cables require twisting

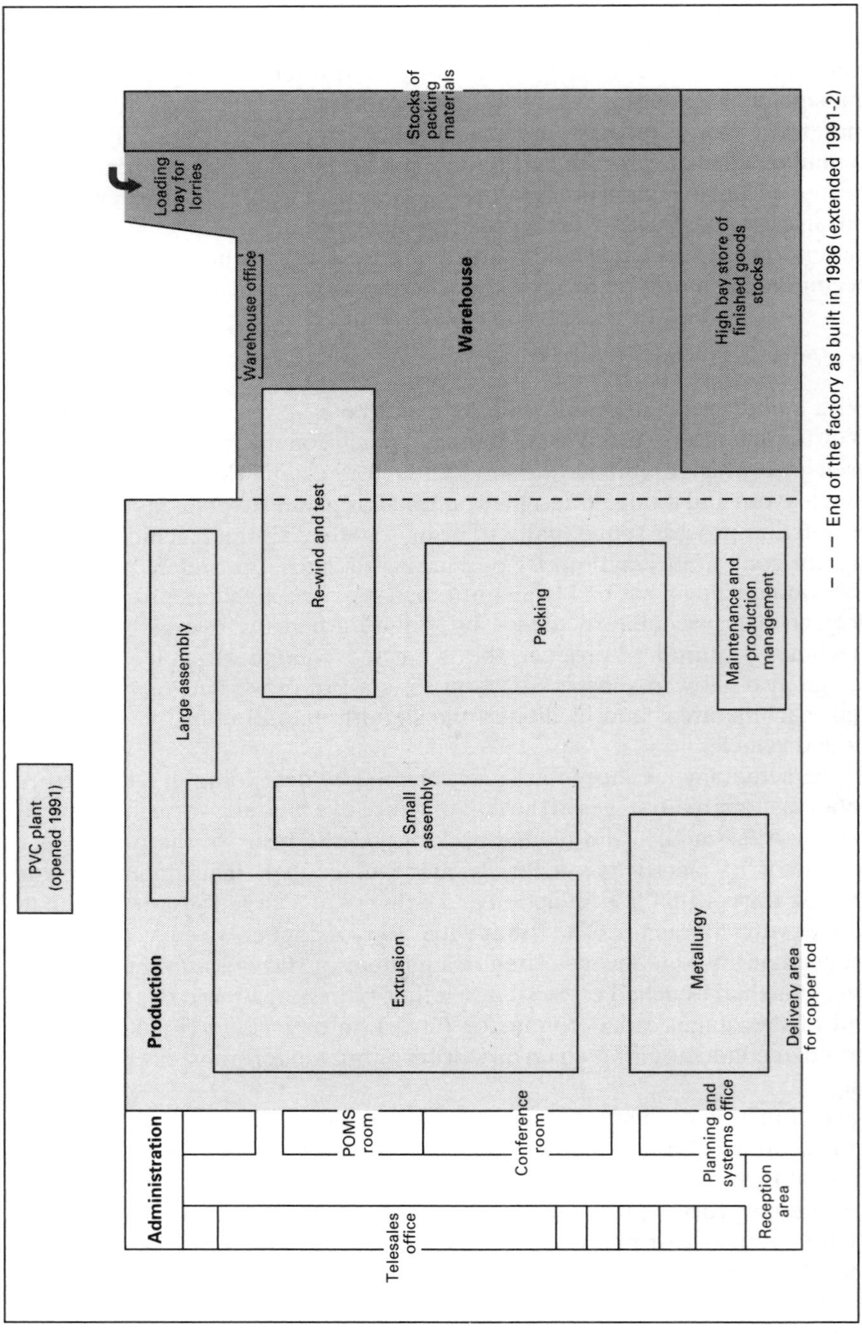
PVC plant
(opened 1991)
Administration
Telesales office
POMS room
Conference room
Planning and systems office
Reception area
Production
Large assembly
Extrusion
Small assembly
Metallurgy
Delivery area for copper rod
Re-wind and test
Packing
Maintenance and production management
Warehouse office
Loading bay for lorries
Stocks of packing materials
Warehouse
High bay store of finished goods stocks
- - - End of the factory as built in 1986 (extended 1991-2)

**Figure 4.1** *Zonal layout of the Aberdare factory, 1995*

together (like making ropes) before they are ready for packaging and dispatch, and so need to be assembled by a 'laying up' or *small assembly* machine. Other types of general wiring, particularly the heavier small industrial cables, require 'armoured' protection through a covering of galvanized steel wire. This cable goes through insulation and laying up as in the previous example, but then goes back to the extruder for a 'bedding' sheath before being transported to the armouring machine (*large assembly*). This twists around 30 fine steel wires round the sheath in a continuous process until it forms the right depth of armour. It is then taken back to the extruder so that a final jacket can be put on the armour, whence it goes to a final test machine before being re-wound and cut to the right length for dispatch to the customer.

More simple and basic cables can go straight from extrusion to the *rewind and packing* area, where they are given a final test before being cut into the required length and rewound onto the appropriate size drum (for larger orders) or put in a cardboard box (for smaller lengths) for dispatch and sale. The majority of building wires (in quantity, but not in price or profit) end up as 100 m lengths in cardboard cartons for use by electrical contractors and electricians in various types of house and building wiring.

The project team decided to lay out the Aberdare factory logically in four main zones corresponding to the four main stages of manufacture (see Figure 4.1). The main connecting points between each zone were then placed as close as possible to each other, so that, for example, the extrusion line for heavier industrial cable finished right next to the large assembly machine, which puts the steel-armoured sheath on the cable prior to final insulation. Within each zone the equipment was arranged sequentially in the order of production: for example, the initial 8 mm rod breakdown machine was placed next to the multiwire machines in the metallurgy section.

Wolstenholme created a systems architecture which was also modular. Each machine was designed to carry out a range of self-contained tasks (allowing flexibility in case of machine breakdown), and varying amounts of space were left between them so that, if capacity needed to be increased in future, another machine could be slotted in. The exact number of machines for each zone was decided on the basis of the 10-year market forecasts of the 1984 Roberts report, slightly expanded to meet the increased market share demanded by the board of directors in 1985 as a condition of investment. In the event, there were seven machines in metallurgy, five in extrusion, five in small and large assembly, and seven in rewind and packing, including final test machines. This was just a half of the equivalent number in the old Southampton general wiring factory, but with a 50 per cent higher capacity.

## The Heart of the CIM System

Stokes's original formulation of project objectives in 1984 had presented the project team with three main guidelines: to be imaginative and experimental;

to devise a system which covered all stages of the business process from raw material purchase to the banking of customer payments; and to devise new methods – or confirm existing ones – with the aim of minimizing costs and maximizing return on investment. By late 1985 the priority had moved decisively in the direction of devising new methods rather than confirming old ones. In the first days after he took up his new job, David Regan had been told in no uncertain terms by the then head of PG's corporate technical department: 'sector wants the project to be as automated as possible'. By late 1985, this view had gathered momentum to the point that all members of the project team were, in Regan's later words (interview, 11 July 1990), 'totally committed to full automation and reducing the people element to a minimum'. This was not the product of any positive strategy to eliminate human beings from the production process. It was derived from their interpretation of what sector managers wanted *and* because, as engineers, they became fascinated by the technical challenge of full automation. They had also become convinced by the trade press and equipment suppliers that full automation was technically feasible. As Regan said later (interview, 11 July 1990): 'I believed in magic, and the technologists said it was possible'. (In a written comment at this point on the draft manuscript he wrote in April 1994: 'And I still do!')

Soon after Wolstenholme was seconded to the Aberdare project team, he, Peters and Regan decided they needed to talk systematically with a number of software and systems consultants (including SEIAF) to find out what level of systems integration for the new factory was technically possible. The discussions between the 'systems' subgroup and various software houses differed in important respects from those of the technical subgroup with the machine suppliers. No plant management systems (level 2 of the CIM hierarchy) for computer-integrated cable production existed at the time. This meant that the level of originality required for PMS software design and implementation was high, with the emphasis very much on 'invention'. In contrast, since many of the machine innovations desired by the Aberdare team were already in operation somewhere in the cable sector, process innovation – machine and process control automation – was a much more balanced mixture of 'invention', 'adaptation' and 'borrowing' (see the introduction to this chapter).

On top of this, most of the machine suppliers had a long track record of successful collaboration with Pirelli and the cable industry, whereas the software companies were either new (as in the case of SEIAF) or new to cable-making. Finally, while the machine suppliers were prepared and able to engage in two-way discussion with the project team about the level of experimentation which was possible in the process control area, this was less possible in the PMS area. Here, meetings tended to consist of the project team saying what they would like and the systems houses responding by saying they thought they could achieve it. And no one in the project team could really contradict them. As David Regan said later (interview, 11 July 1990): 'I would ask them to do things and had no idea whether it was possible, how much work would be involved, whether it would require one hour or one year of software development'.

As a result, while the contract specifications for the various machine systems were fairly precise, the specification for the PMS was not only highly ambitious – in terms of the level of computer integration – but very general, establishing broad objectives but rarely getting down to design detail. In some senses this was inevitable, since the PMS specification was drawn up at the same time as the process control specification and no one knew precisely what operating methods were going to be used. It was against this background that Andrew Slater, head of PG's information systems department, began to express his doubts about the advisability of the full CIM (let alone CIB) strategy. However, when he was excluded from participation in the choice of tendering companies for the PMS, he decided to withdraw from any involvement in this aspect of the project and concentrate on developing the manufacturing and business management systems (the top two levels of the CIM hierarchy).

In the spring of 1986, Mancini, Regan, Wolstenholme and Balandi decided to invite three companies to tender for the PMS systems software contract. Two were located in Britain – one US-owned, the other British-owned. The third (SEIAF) was, as we have seen, Italian–American and located in Italy. SEIAF clearly had many advantages. It had the direct involvement of IBM computer specialists (IBM hardware and software were the standard for the Pirelli group, Pirelli General and the Aberdare project). Its leading members had been involved in months of discussions with members of the Aberdare project team about the systems requirements for the new factory, and they were committed to the ideals of computer-integrated manufacture. Their tender was also the lowest. On the other hand, one of the other tendering companies had also been in discussions with the project team for over a year, and both the others were UK-based and thus geographically close enough to deal swiftly with design and implementation problems as they arose.

The decision as to which company to select was taken in Milan. To no one's surprise SEIAF was chosen. It was SEIAF's first major contract since formally beginning operations in January 1986 (they had been incorporated as a company in September 1985). Over the following two years they spent most of their time working on the Aberdare project.

# Outlining the CIM Philosophy and Objectives at Aberdare

When Arthur Stokes first talked in May 1984 about the need for an experimental approach to the use of computers in cable manufacture, he was talking about a potential rather than a reality, about a general idea rather than details. Between 1985 and 1987 individuals at different levels within the group had worked to translate the idea into a reality for the Aberdare factory. They held extensive discussions with equipment manufacturers and hardware and software suppliers, visited trade exhibitions and conferences, and assimilated new ideas from the engineering trade press. By the summer of 1986 they had also placed orders for new machines with equipment suppliers from four European countries, and engaged an Italian–American systems house to develop programmes for system integration. Separate work was proceeding in PG's corporate information systems department on developing the manufacturing and business management systems software. The essence of what was understood by computer-integrated manufacture remained broadly stable for the next two years.

The general philosophy and approach was summarized succinctly in a 19-page booklet, *Pirelli at Aberdare: Computer Integrated Manufacturing of Cables*. It was written largely by David Wolstenholme and published in mid-1988 to coincide with the technical launch of the project at a London press conference (Pirelli General, 1988). It summarizes clearly the core elements of the CIM philosophy and the main objectives of the project team as they had evolved by that time.

'Put simply', the booklet said, 'CIM means that all aspects of the business are managed by a highly sophisticated "suite" of integrated computer systems' (1988: 5). Elsewhere, CIM was defined as 'the use of integrated computer systems to plan and direct every phase of the business, . . . representing an unbroken chain of information and instructions flowing both ways and involving at any moment of operation 120 million movements of information per second managed by 4000 megabytes of disc storage' (1988: 3, 7). The systems architecture was described in terms of the accepted four-level hierarchy:

- *Business Management System (BMS)*
  - to handle customer orders (entering the system either directly or through telephone connections with the regional distribution depots)
- *Manufacturing Management System (MMS)*
  - to receive the orders from the business management system
  - to control production data; raw materials inventory; production scheduling; acquisition of materials; technical data analysis; costing; purchasing; maintenance
- *Plant Operations Management System (POMS)*
  - to manage all aspects of shopfloor planning (sequencing of machines; inventory management; materials handling system; performance monitoring; quality management)
- *Process Control System (PCS)*
  - to manage and monitor the basic functions of the physical production process (machines, devices)

If we compare this characterization of the Aberdare CIM system with the ideal model outlined earlier on p. 70, it is clear that a computer-integrated business was far from being a reality at Aberdare in 1988. The most interesting omission occurs under the heading of the business management system. In the

1988 Aberdare scheme, the business system was described simply as handling customer orders, with no mention at all of computer integration of the other main business processes: warehouse and distribution, financial and management accounts, sales and marketing, personnel. What was being described in the 1988 booklet was, in the literal sense of the word, a computer-integrated manufacturing system, with production planned by the manufacturing management system, sequenced and coordinated by the plant operations management system (POMS) and controlled by the process control system. And yet the language of the booklet suggested at many points that a full computer-integrated business was not only the ultimate aim but the reality.

## A comparison of traditional and automated manufacture

Whether CIB or just CIM, the intended level of automation of the manufacturing process was extremely high. This can be best illustrated by comparing traditional general wiring cable manufacture at the Southampton factory with what was planned for Aberdare. At Southampton, cables were manufactured in long runs through a large number of separate machine operations. The aim was to achieve economies of scale and to reduce to a minimum the time delays associated with having to change fom one product to another, i.e. altering machine set-ups and supplying new materials. Production was highly inflexible, making it necessary to hold large and expensive buffer stocks of each cable type so that the whole product range would be available at all times. Production planning and sequencing – the size, type and quantity of product to be produced at a specific time on a particular machine – were determined by managerial judgement based on a forecast of future demand (usually derived from past sales patterns) and regular, time-consuming, physical inspections of stock levels to prevent supplies of any one product getting too low.

At Aberdare, the vision of computer-integrated cable production was quite different. Customer orders indicating the degree of urgency and date required would be received and inputted into an IBM System/36 minicomputer. The manufacturing management system (a modified version of IBM software) would then prepare a production plan and define the operations for each machine during a predefined period of time, such as a shift or a day. Computer control would then pass to the plant operations management system (POMS) and one of its two key software elements, the shopfloor supervisor (SFS). The SFS would be in overall control of factory operations. On receiving the list of operations, it would rearrange it automatically according to the actual status of the plant, and then wait for a machine to complete its current task.

Whenever a machine was about to become idle, the SFS would select a new operation from the list, trying to optimize usage of materials and minimize the time required to set up the machine. Eventually all the resources required to carry out the operation would be selected and reserved for it. To do this the SFS would be constantly aware of the current status of the machines, the positions of the bobbins and their content, the raw materials availability, and what the operators were doing.

At this point control would pass to the second main software element of POMS, the machine supervisor (MS). The MS would control the operations on each individual machine, telling the machine plcs how a particular operation was to be executed and instructing operators if manual intervention was required. It would also issue requests to the automatic guided vehicles to carry bobbins to the machine or take them away. While all these operations were being carried out, the MS would constantly monitor the performance of each machine, collecting data about the quality of the cable and the length produced. It would also take appropriate action to deal with abnormal situations when these were indicated by automatic alarms. The MS would notify the SFS of all relevant events, and provide a full report on each operation as soon as it was completed. In turn, the SFS would continually send reports to the manufacturing management system on the tasks which had been carried out, thus closing the information flow.

In sum, the Aberdare system envisaged an integrated set of computer links between the first three levels of the CIM hierarchy, coordinated by POMS. As Wolstenholme described it:

> The heart of the system, POMS, continuously controls the plant processes and sequences the production minute by minute. Real-time data from all company resources – availability of materials, machine status, employee skill levels, work-in-progress, inventory management parameters – are used to select, on an event driven basis, the job each machine is to do next and the sequence of movements of material about the production zones. There are no unplanned activities, no dead stocks of 'waiting' material, and no over-booked or under-booked machines or operations. In short, CIM provides Time and Motion efficiency of an unprecedented order. (Pirelli General, 1988: 7)

This statement makes it clear that POMS was not only regarded as a *management information system*, but also as a *computerized system of management*. It was seen as the central mechanism for managing the whole process of manufacture from order input to completion of the finished product. This was to include: the sequencing of the order in which products were to be made (according to priority of customer demand); preparation and just-in-time delivery of raw materials for each stage; instructions to set up, control and finish each machine process; just-in-time transfer of bobbins at each stage of the process; continuous quality management; and an 'automated paperless system' of 'immediate management accounting information' from the system's 'precise knowledge of product location, status and unit cost' (Pirelli General, 1988: 13). However, there was no mention of stock control, distribution, or customer billing tasks.

POMS was thus intended to carry out automatically many of the routine tasks usually performed by managers and operators. It was no coincidence that the two basic software elements within POMS were called the 'shopfloor supervisor' and 'machine supervisor', corresponding in the traditional factory to the shift supervisor and machine operator respectively. The shift managers were to be relieved by POMS of their traditional tasks of controlling events on the shopfloor, enabling them to concentrate on the overall coordination of resources (human, material and financial) and the management of exceptional situations (Trezza, 1988). Operators, too, would no longer need to deal with the normal routine of machine operations, but would monitor the status of production and manage faults and quality problems as they occurred. POMS was at the heart of the vision splendid of a computer-driven automated factory.

## Presenting the Project to the Technical Media, July 1988

It was with the ideal of a computer-automated factory that Pirelli captivated the technical media at its London press conference on 12 July 1988. The conference was opened by Lord Westgate, chairman of the PG board of directors. He then introduced an eight-minute video on the aims of CIM at Aberdare which had been specially commissioned for the occasion. This was followed by three 10-minute presentations, by Arthur Stokes on the philosophy of the project, David Regan on implementation, and David Wolstenholme on POMS. The conference was concluded by short presentations from Arturo Giodotti (marketing and sales manager, SEIAF) and Peter Smith (marketing development manager, IBK (UK)), who spoke about the wider benefits of CIM.

Much of the trade press – from *Electronics Times, Electrical Products, Electrical Review*, and *Electrical Contractor* to *Plastics and Rubber Weekly* – celebrated Pirelli Aberdare as the factory of the twenty-first century. The *Financial Times* (19 July 1988) headed its report, 'Computers in control', describing the plant as the first comprehensive CIM system in the European cable industry and 'a pilot for the Pirelli group as whole'.

Most of the publicity surrounding the opening of the factory concentrated on the wonders of computer automation and particularly the AGVs. However, some commentators noticed that the project also involved a new philosophy towards people, with an emphasis on a flat organization structure, teamwork, flexibility, single status employment and high levels of employee participation. On page 7 of the company's booklet on CIM at Aberdare (Pirelli General, 1988), POMS was described as the heart of the CIM system. Just eight pages later, it said: 'at the heart of the new Aberdare plant are its people'. Peter Large, an experienced technology journalist, picked up the wider significance of Pirelli's CIM experiment – the link between technical and human resource innovation:

> Propaganda has become reality on the outskirts of Aberdare. A . . . factory that at first blush looks utterly boring – it makes household electrical wires – is Britain's (and perhaps Europe's) best example yet of twin themes that have been preached for more than 20 years but never realized. The first theme is routine total automation: the second, multi-skill employment . . . (*Guardian*, 4 August 1988)

Large's article was headed: 'Pirelli slips into total automation on its Welsh industrial kibbutz'.

# Summary

In this chapter we have traced the way in which the computer-integrated manufacturing system for the Aberdare project was designed between 1984 and 1988. The original plan to have a sector-wide steering committee overseeing the whole project was gradually superseded by the creation of a number of more specialized groups focusing on particular areas. In the initial stages the main focus was on level 1 of the CIM hierarchy, the machine and process control system. The 'technical' project team was able to learn important lessons from the experiences of other factories within the Pirelli group and other UK investors in advanced manufacturing technology. In one machine area – extrusion – they recommended an extremely high level of machine and process innovation, with important consequences, as we shall see. They also failed to resolve who should bear responsibility for the interfaces – the electronic handshake – between the various machine systems and the plant management system.

The 'systems' project team designed a highly robust and logical systems architecture and factory layout. However, no one in the team was a specialist in software design, and the specification of the plant management system was highly general, ambitious and experimental. The contract for the PMS was awarded to a systems house for whom this was the first major contract.

While there was regular contact between the 'technical' and 'systems' project teams, the development of the two highest levels in the CIM hierarchy – the manufacturing and business management systems – was left largely to members of PG's corporate information systems department. The BMS in particular, which covered areas such as sales orders, billing and accounts, was subject to stronger degrees of corporate control and coordination than the other three areas, and the degree of project management autonomy and integration was much more limited.

The CIM philosophy and objectives, as outlined in a booklet in mid-1988 for the technical launch of the project at a London press conference, were described in terms of a computer-integrated business, although most of the emphasis of the pamphlet was on the manufacturing side of computer integration. Compared with traditional methods of cable manufacture, however, what was being proposed represented a major strategic innovation, both in technology and in the role of supervisors and production operatives.

# Questions for Discussion

1 Evaluate the company's main objectives in choosing a computer-integrated manufacturing system.

2 Discuss the main features of (a) machine automation and (b) computerized management systems chosen by the project team in 1985–6.

3 To what extent and why did the CIM system chosen by Pirelli deviate from the typical CIM system structure outlined in the chapter introduction?

4 Evaluate the main strengths and weaknesses of the project management arrangements for designing and choosing the various elements of the CIM system.

5 To what extent is it accurate to describe the eventual system design as a computer-integrated business rather than a computer-integrated manufacturing system?

# 5

# The Retreat from Full Automation

## Thematic Focus: The Implementation of Technical Innovation

In Chapter 1 we noted that the essence of strategy is 'the deliberate and conscious articulation of a direction' (Kanter, 1983: 294). We also suggested that strategic change is rarely the result of a linear process of decision-making in which strategy formulation (what to do) is followed sequentially by strategy implementation (how to do it). In practice, strategies are not fixed at some single decision point; they are evolved and elaborated continually as the internal and external environments of organizations change. Successful strategic change involves both the setting of a clear and conscious direction and an organizational capacity to adapt strategy to continually changing and unpredictable circumstances. Finally, we noted that experimental and highly innovative strategies can have a galvanizing effect in solving fundamental organizational problems, but they also require high levels of employee trust and risk-taking and entail a greater possibility of failure as well as success.

In the first two chapters we looked at the process of setting broad strategic objectives and the crucial role of the senior executives with the organizational power to mobilize support for strategic change. In Chapters 3 and 4 we examined two issues – site location and technology – on which detailed work still needed to be done to put flesh on the bones of the strategy (Chapters 6 and 8 will look at parallel developments in human resource management and industrial relations). If implementation is defined as a 'cascade process of ever more detailed decisions which occur in the context of primary strategic aims' (Pettigrew and Whipp, 1991: 175, following Hrebeniak and Joyce, 1984), then we were already concerned in these chapters with the early stages of implementing strategic change. In this chapter, we will be concentrating on the first three years of the implementation of technical change in the more conventional use of the term – the installation, commissioning and initial operation of new technical systems. Before we do this, we will review some of the central findings from the literature on implementation with particular reference to advanced technical change. Further discussion of the general literature on the implementation of change can be found in Chapter 7.

The initial implementation of technical change can be broken down for analytical purposes into two main stages: *installation*, the delivery and assembly of new systems in their designated location and their connection to the appropriate networks and sources of power; and *commissioning* or 'de-bugging', running and testing the new systems in a simulated operation and ironing out major faults or 'bugs' prior to proper operation (see Clark et al., 1988: 167–77; Greenwood, 1988: 218–36). With computer-integrated or flexible manufacturing systems, of course, it is not only individual pieces of hardware and software that must be installed and tested, but also the effectiveness of the links between them. Modifications made to individual parts of the system during these stages will almost certainly have knock-on effects for the functioning of the system as a whole.

According to Greenwood, one of the leading experts in implementing CIM systems, 'probably the single most important factor to be borne in mind when considering installation and commissioning . . . is that this process is likely to be both lengthy and resource-intensive' (1988: 218). It is also extremely difficult to estimate how long commissioning will take, as this depends on the newness and complexity of the system, how many problems are exposed during early testing and, once exposed, how easy they are to correct. He concludes: 'it is during this phase of the project that one finds out how comprehensive the detailed design of the system really is' (1988: 218). On all these counts – system complexity, machine problems, problems with detailed design and specification – PG experienced difficulties at its new Aberdare site.

There are three main ways in which a computer-integrated or flexible manufacturing system can be implemented (see Greenwood, 1988: 220). First there is what Greenwood calls a *greenfield site approach*, in which the entire system – machines and machine controls, plant, manufacturing and business management systems (see Chapter 4) – is installed and commissioned simultaneously. Second, the *linked islands of automation* approach involves installing the system in islands or zones, which can then be linked together at a later stage. Finally, there is *piecemeal evolution*, the installation of plant and equipment piece by piece, eventually forming islands of automation, which can then be gradually integrated into a fully-fledged CIM or FMS system. On Greenwood's analysis the linked islands approach is the most practical method of progressing towards full automation. He counsels strongly against moving directly from a traditional production environment to full automation, 'since the strain placed on both the shopfloor personnel and production control systems might be quite unreasonable' (1988: 220).

US research by Alexander (1986) has identified the most frequent problems encountered by companies when implementing strategic change such as large-scale technical innovations. Alexander began his study by identifying 22 potential implementation problems from the research literature. He then asked senior managers in 93 medium- and large-sized US companies to choose a strategic decision in which they had recently been involved and to identify which of these problems had occurred. The 10 which occurred most frequently are listed in Table 5.1. He then went on to ask his

**Table 5.1** *Ten most frequent strategy implementation problems*

| Problem | Frequency of problem (% of 93 companies surveyed) |
|---|---|
| 1 Took more time than originally allocated | 76 |
| 2 Major problems arose that had not been identified beforehand | 74 |
| 3 Ineffective coordination of implementation | 66 |
| 4 Competing activities and crises distracted attention | 64 |
| 5 Inadequate employee capabilities | 63 |
| 6 Inadequate training and instruction given to lower level employees | 62 |
| 7 Adverse impact of uncontrollable factors in the external environment | 60 |
| 8 Inadequate leadership and direction by departmental managers | 59 |
| 9 Lack of detailed definition of key implementation tasks and activities | 56 |
| 10 Inadequate information systems to monitor implementation | 56 |

*Source*: Alexander, 1986: 252

**Table 5.2** *Six most highly rated implementation problems for low success companies*

| Problem | Mean rating (1 = no problem, 5 = major problem) |
|---|---|
| 1 Took more time than originally allocated | 3.23 |
| 2 Major problems arose that had not been identified beforehand | 3.19 |
| 3 Adverse impact of uncontrollable factors in the external environment | 2.87 |
| 4 Lack of detailed definition of key implementation tasks and activities | 2.81 |
| 5 Competing activities and crises distracted attention | 2.74 |
| 6 Inadequate leadership and direction by departmental managers | 2.71 |

*Source*: Alexander, 1986: 257

respondents to rate these problems as minor, moderate, substantial or major. He also grouped the firms into high, medium and low success depending on the relative degree of success (based on the extent to which they had achieved their initial objectives within budget) in implementing the particular strategic decision. He found that the presence of the higher rated implementation problems were closely correlated with low rates of success. Six problems stood out as particularly crucial (see Table 5.2). The two most highly rated problems for low success companies – that implementation takes longer than originally planned, and that major problems emerge that had not previously been anticipated – were only too apparent in the first years of implementation of technical change in the new Aberdare factory.

---

# From Empty Shell to Official Opening

### Project management

In March 1987 David Regan formally took over the new Aberdare factory building. At this stage, it was not much more than a shell. It had no machines, no computers, no communications network, and no people, just a few telephone lines. Project implementation in the period until the official opening in July 1988 was dominated by three parallel processes: recruitment; the initial training of staff (both to be discussed in Chapter 7); and the installation, commissioning and initial operation of new machines.

During this 16-month period, Pirelli General as a company needed to ensure the continued supply of general wiring cable to its existing customers. In particular it needed to coordinate the gradual phasing out of the old Southampton plant as the new Aberdare factory came on stream. The plan as of March 1987 was for the Southampton facility to run down during 1987 and 1988 and to close by Christmas 1988. In terms of project management, Regan and his operations manager Martin Dawson (of whom more later) were to assume responsibility for production during the changeover period, while Ian Scott, the divisional manager of the Southampton factory, would retain control of sales, marketing, customer relations, and management accounts. As a result, the Aberdare commercial manager and site accountant spent much of 1987 and 1988 in Southampton rather than Aberdare.

This division of labour was understandable. However, it did mean that during 1987 and 1988 Regan and his operations manager Dawson – who was in charge of production, maintenance and systems engineering in the new factory – spent most of their time concentrating on production and systems issues, while commercial and financial matters tended to be neglected. This mirrored the strong emphasis on the first two levels in the CIM hierarchy in the design of the new system at Aberdare, and the relative neglect of the business management system.

## Installation and commissioning – an extended honeymoon?

According to Greenwood, as we have seen, there are three main ways of implementing a CIM system: greenfield site, linked islands of automation, and piecemeal evolution. Installation and commissioning at Pirelli Aberdare was a mix of all three approaches. The initial plan was for the metallurgy equipment to be installed and commissioned first, machine by machine, eventually forming an island of automation capable of supplying the Southampton factory with finished strand for extrusion, assembly and packing (this is in fact what happened from the spring of 1988). This was to be followed by the installation and commissioning of machines in the small assembly, extrusion and large assembly areas. The test, rewind and packing equipment would then be transferred from Southampton to Aberdare towards the end of the implementation process in early 1988. Once the commissioning of machines in various islands of automation was complete and they were performing reliably, it was intended to begin trials to integrate them with one another through the computer-controlled plant operations management system (POMS).

Project management of machine implementation was split between Aberdare, PG corporate, and Pirelli sector technical specialists, with members of the PG corporate technical department playing the dominant role. This division of labour reflected in part the relative interests and influence of these groups in the specification, design, and ordering of the various components of the CIM system. In the event, due to recruitment and other pressures (see Chapter 7), the only Aberdare employees directly involved in installation and commissioning of the metallurgy machines were three production engineers, one systems engineer, and a small number of production staff.

Apart from their initial role in late 1984 and early 1985, sector managers in Milan had had little involvement in the development of the plant management system or in wider system integration (although they did play a decisive part in the decision to select SEIAF as systems designers for the project; see Chapter 4). Andrew Slater, head of PG's information systems department (ISD), in contrast, was initially very interested in these areas, but had effectively opted out of involvement in POMS when he was excluded from the decision about the choice of systems designers in 1986. From then on, ISD concentrated on the piecemeal evolution of various parts of the manufacturing management and business management systems. As a result, the detailed design and implementation of POMS, and liaison with SEIAF, was left up to the two main members of the Aberdare project team seconded from corporate HQ – David Wolstenholme and Gordon Peters – with the assistance of two young systems engineers on the Aberdare payroll. During 1987 they held regular – sometimes week-long – meetings with SEIAF staff to discuss the detail of POMS specification and development. The meetings usually took place in SEIAF's education centre situated above the coastal town of Genoa.

Throughout 1987, the processes of machine installation and commissioning and the development of POMS software proceeded in almost complete separation from each other. This led to an increasing fragmentation of overall project management. It was part of Regan's management style to motivate individuals by giving them a large amount of freedom and discretion to carry out their work. He was able to achieve remarkable levels of commitment and motivation through this approach. However, as we shall see, it also led in 1987 and early 1988 to a sense of drift and fragmentation – what one of his middle management staff later called an 'extended honeymoon'.

In the summer of 1987, German engineers arrived to install the first machine in the manufacturing process, the rod breakdown machine. They were soon followed by more German engineers (from a different company) to install machines in the rest of the metallurgy section. In each case, machine installation and commissioning specialists from the PG corporate technical department, together with one Aberdare production engineer and a small hand-picked group of Aberdare production staff, were assigned to work alongside them. In so doing, the Aberdare staff were able to gain valuable knowledge about the intricacies of the new metallurgy machines, as well as keeping an eye on the installation and commissioning process. By the end of 1987, machine installation in the extrusion and assembly areas had begun, and some of the packing equipment had started to arrive from Southampton. By then, too, good progress had been made laying the underground optical fibre ring system which was to provide the links between the various parts of the CIM system, as well as the internal communications infrastructure on the site. By the spring of 1988, the metallurgy machines, while by no means fully operational, were able to produce sufficient finished strand to meet the needs of the Southampton factory, and by the summer the AGVs had arrived and could be activated manually as 'standalone' material conveyors.

Meanwhile, as the machinery was being installed and commissioned, SEIAF engineers began to make regular visits to Aberdare. However, as soon as they started trying out POMS in the metallurgy section, it became clear that the software requirements were far more complex than had been originally envisaged. On top of this, some parts of the machines in metallurgy and extrusion had been modified during installation and commissioning, mainly because of design faults and faulty machine specifications, but also because of the desire of PG's corporate technical specialists to enhance machine performance. Each time this happened, SEIAF had to go back to the drawing board to modify the POMS software to take account of the machine modifications. All of this work was, of course, additional to the original contract. By mid-1988 the 'system cost', budgeted in 1985 at around £1 million, had escalated to over £3 million (Pirelli General,

1988: 5). This combination of basic mechanical problems and lack of POMS implementation meant that, at the time of the formal opening of the plant by Prince Charles in July 1988, it was still not possible to produce a complete cable from start to finish, only individual elements of strand.

### The official opening by Prince Charles

It seemed natural for Prince Charles to be invited to open the plant, and natural for him to accept. For many years he had shown a strong interest in the development of UK manufacturing. He was also Prince of Wales, where a large amount of recent UK manufacturing investment was concentrated. Indeed, he had played an important role in initiating the growth of inward investment to Wales by Japanese companies. In 1970, during a visit to Expo '70 in Japan, he was introduced to Akio Morita, co-founder and chairman of Sony, who had just announced his company's new global production strategy. It was not long before the Prince asked Morita if Sony had any intention of building a plant in the UK. Morita said that the company did not have any such plans as yet, and the Prince replied that if Sony were to build a factory in the UK, Sony should not forget Wales (Morita, 1994: 131). Within a year the decision was made, and in 1973 Prince Charles formally opened Sony's colour television factory in Bridgend, South Wales. By the late 1980s it was employing over 2000 employees and its success had played a major role in encouraging other Japanese firms to locate in Wales.

In the early afternoon of 22 July 1988, the royal helicopter landed on the new car parking area outside the factory. Prince Charles was met by the two main strategists behind the new factory, Arthur Stokes, now chairman of Pirelli General, and David Regan, manager of the new unit. When the visiting party of invited guests – local dignatories, members of the national and local press, members of the PG board and corporate departments, and representatives from group headquarters in Milan – entered the factory it was spotless and humming, with not a commissioning engineer in sight. The whole factory and its staff were a buzz of activity. The AGVs were gliding busily about, albeit to no real purpose and under invisible control. The cable-makers present soon realized that no finished cable was actually being produced on the day. However, the local dignatories and national press were clearly oblivious to this, totally convinced that what they were seeing was the automated factory of the future. Impression management won the day. All the staff played their part. Regan, in his inimitable style, described the whole event as 'shit hot'. The opening was a major success.

## The End of the Honeymoon

The success of the official opening could not hide the fact that the factory was still not producing a complete cable. And it looked unlikely to do so in the foreseeable future. Between 1985 and mid-1988 the experiment had enjoyed the strong support of sector management in Milan and the PG managing director. As a result, Regan and his senior management team had been given most of the resources they had asked for: buildings, machines, software, information technology and computing, and also staffing and training (see Chapters 6 and 7). Because of the experimental nature of the project, they had also been sheltered from the traditional disciplines of management accounting and given what appeared to be a generous amount of time to get the new plant up and running. There was no customer or production pressure, which was still concentrated on the Southampton factory. This relatively 'easy ride' in the initial period of implementation generated a mixture of envy and disbelief among corporate managers in Southampton, but they believed little could be done given the strong support of their MD and of sector management for the project.

By the end of 1988, 21 months after the factory had been first occupied, commissioning of some of the machines was still not complete. Many of those that were in full operation were still breaking down regularly or producing faulty strand. During 1988 the factory never once reached anything like its management plan output target. The planned closure of the Southampton factory in December 1988 had to be postponed for a further year.

## The appointment of an internal troubleshooter

By this time, sector managers in Milan began applying strong pressure on Mancini to rectify the situation. Regan was felt to be struggling unsuccessfully with a whole range of problems, always promising that things would soon come right, but never seeming to turn the corner. In the view of sector management, there was a general malaise among the senior management team at Aberdare. They had somehow lost their way and appeared frightened of the extent and depth of the problems they faced. Sector's medium-term solution to the problem was not announced until later in the spring. Their short-term solution was to persuade Mancini to second one of his senior corporate managers to act as a troubleshooter to sort out the main outstanding technical problems. He would have a completely free hand to go to Aberdare, decide on what was wrong, and take remedial action to achieve the required levels of production.

At the end of January 1989, Stephen Cole began what turned out to be a four-month, four-days-a-week stint at Aberdare. Cole had joined PG as a graduate in 1968 and, after a number of senior production jobs, was appointed in 1987 to a new corporate position in charge of 'organization, systems and structures' with a direct line to the managing director. His main task was to look at PG organization structure and find ways of streamlining it.

Cole spent the first few days of his new assignment talking at length to Regan, operations manager Martin Dawson and the three shift managers. He then spent nearly two weeks out on the factory floor listening to production and maintenance staff and observing what was happening. By the end of February, he felt in a position to make an initial assessment. First, he identified major problems in production management. With one exception, he felt that the shift managers were too remote from the problems of the factory floor, they had minimal technical support on shift to deal with recurrent problems, and there was little or no coordination between them (they were operating a 3 × 8 hour continuous shift system with only a small amount of overlap). Some of the operators had not yet been trained properly on the machines (see Chapter 7). On top of this, Regan was spending a large amount of time on external matters such as discussions and negotiations with customers and management meetings in Southampton, while the operations manager Martin Dawson was concentrating most of his time and energy on POMS implementation, which was massively behind schedule. No one appeared to be in overall control of the production and the machine problems. The result was that managers and staff were spending much of their time fire-fighting and devising short-term solutions to try and keep the machines running.

He proposed two immediate solutions: the establishment of a new position of manufacturing manager – located between the shift managers and the operations manager – to take overall control of production and coordinate the activity of the three shift managers; and the immediate appointment of two shift production engineers (two of the original three had left the site in mid-1988 and had not been replaced) so that each shift would have specialist technical support permanently on hand to deal with recurrent machine problems.

Cole had full authority from Mancini and sector management to force these recommendations through. However, he wanted if possible to try and persuade Regan and Dawson – the latter was a personal friend of long standing – to accept his proposals before wielding the big stick. While they readily agreed to the proposal for two additional production engineers, they were strongly resistant to the idea of creating a new level of production management. They had argued long and hard about this issue in 1986–7 (see Chapter 6) and concluded that short hierarchies were essential to generate staff commitment and motivate them to take responsibility for the day-to-day operation of the factory.

However, after long discussions, Regan and Dawson were persuaded, by one argument in particular. The original organization structure had been designed on the assumption that once the new machines and software systems had been commissioned, the factory would run automatically without the need for a great deal of human intervention. In practice, as we have seen, the new equipment had been dogged by all kinds of problems, and the heart of the CIM system, POMS, was still not up and running. The original organization structure was simply not appropriate to the task in hand. As a 'temporary measure', they agreed to the creation of the new post. They also pressed at the same time to increase the number of skilled technical and systems support staff to tackle the remaining problems.

Reluctantly, the ideals of lean production and a de-layered management were temporarily shelved, indeed reversed, in the interests of effective implementation.

Determined to do everything he could to make the new factory a success, Mancini agreed to all the proposed changes. By August 1989, eight new maintenance staff, two new shift production engineers (a third had been in post since February 1988), and three new posts of automation engineer (electronics-trained maintenance staff graded at management level) were appointed to mount a major attack on the continuing machine problems. And one of the existing shift managers – David Thomas, a Welshman who had previously worked for Delta, one of Pirelli's main competitors – was promoted to the new position of manufacturing manager. This left the other two shift managers with severely bruised egos and what amounted to a downgrading of their jobs (they now reported to Thomas rather than directly to operations manager Dawson).

## Dealing with machine problems on extrusion

Cole's other main preoccupation was to get to grips with the machine problems in the plant. The most important of these was in extrusion. The extrusion section contained the most risky and experimental machine innovation in the whole factory (see Chapter 4). This involved the tandemization of insulation and sheathing in one operation, running three pieces of strand simultaneously through the machine and colouring them at the same time. Extrusion is arguably the core of cable-making. Not only does every cable type have to go through the process, some have to go through it more than once, for insulation, bedding and sheathing. If the extrusion section is not working, then the whole factory is in difficulty. This is exactly what happened at Aberdare in 1988 and 1989 (on my first visit in April 1989, all five extrusion lines were down, a not unusual occurrence at the time).

Technical complexity and commercial confidentiality prevent a full discussion of the causes of the extrusion problems. Suffice it to say that Cole had long discussions with Regan, Dawson, Balandi, and the head of the PG technical department about what to do. In the end, they agreed that since this was one of the most interesting and innovative experiments in the whole plant, and since they had already spent large sums of money on it, they would continue to try and make it work, even if it took more time. Mancini accepted the recommendation, but he did not stay to see it implemented. In the spring of 1989 it was announced that he was to retire as managing director at the end of May.

## Felipe Martinez replaces Mancini as managing director

Mancini's replacement was to be Felipe Martinez, the managing director of Pirelli in Spain. Martinez had spent the first 20 years of his working life with

Pirelli's main cable competitor in Spain, starting as a technician engineer, then taking an engineering degree and moving through a series of middle and senior management positions in production, finance and sales. He was recruited to Pirelli Spain in 1981 as MD, and within four years had turned the company round, partly by a programme of staff reductions (from 1500 to 900 between 1981 and 1985), partly by project managing an extensive upgrade of its previously unprofitable general wiring factory. No one was in any doubt that his appointment was meant to shake up the UK company, and the Aberdare building wires unit in particular.

Martinez's aproach differed in significant ways from that of Mancini. He had no particular philosophical or personal commitment to the Aberdare experiment. He also had a successful alternative experience to draw on, having in the mid-1980s led the Spanish project management team which had installed and commissioned a new but more conventional building wires unit on time and to budget. The machines in Spain were advanced but tried and tested, there was no software-driven plant management system or materials handling, the new plant was located just two miles from the main factory, management and workforce were hand-picked from the best staff in the company, and project management had been tightly managed with little interference from sector. Finally, as the new plant came on stream, Spain began to benefit from a major upsurge in economic activity following its accession to the European Community in 1986. By the late 1980s, in sharp contrast to Aberdare, the Spanish building wires factory was already generating a good level of profits before tax.

Martinez's basic approach to the Aberdare factory was a pragmatic one. It had been, and still was, an interesting experiment. However, it had run into a whole series of delays and difficulties. It was his task to identify the main problems, sort them out, and make the factory profitable by whatever means he could. His immediate aim was to set up procedures which would enable him to monitor the unit's performance, something which neither Mancini nor Regan had done systematically.

To this end, he established a regular monthly 'operations review' meeting at Aberdare with a substantial degree of PG corporate management involvement. Apart from himself (in the chair), it was attended by the PG corporate technical manager (from 1 June 1989, Stephen Cole), the PG corporate information systems department manager (Andrew Slater), and the 10 senior managers and line managers in the Aberdare plant (all the line managers except the shift managers). The agenda consisted of reports, followed by discussion, on the performance of each of the main areas of the factory's operations: production, maintenance, machine performance, planning, systems, sales, purchasing, finance, and personnel. Sometimes the meetings lasted all day, and even went into a second day when they were not able to agree solutions to particularly complex problems. Until this time, Regan – together with the divisional managers of the other main operating divisions – had reported on progress at a regular monthly divisional managers' review meeting in Southampton. Now, the relevant line manager at Aberdare was called to account directly by the MD, and the performance of the Aberdare plant was scrutinized and discussed as the sole agenda item. Minutes were also produced which identified actions arising from each meeting and the individuals responsible.

After only a few months, the force of Martinez's views and new approach had already begun to be felt. Despite the consensus which had emerged from the Cole review, Martinez instructed Regan that the experiment with gear pump technology in extrusion should end in the interests of production. Instead he proposed the installation of an alternative method of PVC coloration known as the 'piggy-back' system (the coloration system was attached to the existing machine like a piggy back). This system had been tried and tested already with great success in the Spanish building wires factory. Installation and commissioning of the first 'piggy-backs', under the project management of a member of PG's corporate technical department, took place in 1990 and they proved so effective that it was soon decided to install the same system on a further two extrusion lines. With consistent development work on the machines throughout 1989 and 1990, and the successful introduction of the piggy-backs in 1990 and 1991, the extrusion section gradually developed into one of the fastest and most reliable parts of the factory.

# The Retreat from Full Automation

The more difficult technical decision to emerge from the 1989 operations review meetings was to retreat from the original ideal that computer systems would plan and direct every phase of the business. In fact, the decision to introduce a greater degree of manual intervention in the production process was a confirmation of a conclusion to which Wolstenholme, Dawson, Peters and Regan had already come, reluctantly, themselves. To understand why, we need to go back to 1986 and trace the main developments in the top three levels of the CIM hierarchy (see Chapter 4): the business management system (BMS), manufacturing management system (MMS), and, above all, the plant operations management system (POMS).

## *The business management system remains outside the Aberdare CIM system*

The BMS at Aberdare covered a number of activities – accounts, sales, purchasing, personnel – which were common across the whole of Pirelli General. It was therefore always going to be difficult to achieve their integration with the rest of the Aberdare system. For example, many customers bought cable products from more than one division of PG, and there were strong grounds for having a common sales order and billing system across the company rather than allowing Aberdare to go it alone. In addition, multidivisional and multinational organizations like the Pirelli group tend to have common financial accounting procedures which enable standardized measurement and central control of financial performance across the group.

These important, if not insurmountable, structural pressures against the development of dedicated single-site commercial and accounting systems were reinforced in 1987–8 by the particular way in which these functions were implemented. It will be remembered that, while responsibility for production was transferred to David Regan in 1987, responsibility for the commercial and accounting functions of the division remained throughout 1987 and 1988 with the divisional manager of the Southampton site. During this period, much of the initial work on computer integration was being done at Aberdare, but the two senior managers responsible for the commercial and accounts functions were based in Southampton, learning the established PG systems rather than developing new ones.

What automation there was in the BMS involved setting up local computer workstations for all office staff, modifying and refining the corporate commercial software so that it could be aligned more with the local systems at Aberdare, and establishing the rudiments of computer-based self-contained personnel and accounts systems. In addition, the installation of a fibre-optic network – which linked all the machines on the shopfloor and in the offices – enabled the creation of an integrated computer-based communication system across the site. By 1990, then, the BMS at Aberdare had a fully-fledged communications infrastructure and a small number of islands of automation, but it was hardly integrated at all with the rest of the Aberdare CIM system.

In the dispatch and distribution area, however, decisions were made in 1989 which had a significant effect on the potential for future integration between the BMS and the rest of the Aberdare CIM system. Until 1989, little change had been made to existing arrangements. Distribution was dominated by the transport of large consignments of finished products – carried by the company's wholly-owned haulage company – to various regional warehouses, where they were unloaded, cut and sorted by PG staff, and then dispatched in small batches to local electrical wholesalers for sale to customers. In 1989 this cumbersome and inefficient system, inherited from the old general wiring division, involved the employment of 50 staff in five separate locations (Bradford, Bristol, Gillingham, Glasgow, and Lichfield), adding something like 7 per cent to the basic production cost of the cable. David Regan had always intended to continue rationalizing the number of warehouses (there had been 10 in 1985; see Figure 3.1). He had also toyed with the idea of having one big warehouse, not at Aberdare, where the transport costs to the rest of the UK were felt to be too high, but somewhere in central England or near London, which could be linked to Aberdare by computer.

Soon after his arrival, Martinez announced that he too was in favour of rationalization of warehousing, dispatch and distribution on one site.

However, he was convinced that part of the success of the Spanish building wires operation was due to the co-location of the warehouse and the factory on the same site. This spatial closeness had also created much greater employee awareness of the links between production operations and the need to meet customer requirements. So, despite the opposition of the whole Aberdare senior management team on logistical and financial grounds, Martinez was able to persuade the PG board of directors and cable sector management not only to create a single high-technology warehouse (the medium-term financial arguments for this were very strong), but also to site it at Aberdare. As we have seen, there was ample space to construct it as an extension of the existing factory (see Figure 4.1).

## The role of the systems engineering department in CIM development

Whereas much of the BMS was still not directly under the control of the Aberdare management team, the manufacturing management (MMS) and plant operations management (POMS) systems were. The original plan was for the small Aberdare systems engineering department to get the MMS (including order processing and bills of materials) up and running, and then gradually to take over the responsibility for POMS from SEIAF and the two main corporate 'systems' secondees (David Wolstenholme from ISD, Gordon Peters from the PG technical department). Because of the high-tech office environment at Aberdare, systems engineers were also intended to provide hardware and software support to sales, accounts and personnel staff. To carry out all these functions, the systems engineering department was originally allocated six staff – a chief systems engineer and five systems engineers.

In the initial stages, the systems engineers spent much of their time installing, modifying and reprogramming various elements of the BMS and MMS software. However, two were allocated to Wolstenholme and Peters to help with POMS implementation and to prepare for the handover from SEIAF. As we have seen, it became clear by mid-1988 that POMS would not only not be delivered on time, but that it still required substantial modification and development work.

The initial response of Regan and Mancini was to employ more systems staff – some on a temporary basis – to try to resolve the systems problems in general and POMS in particular. Between October 1988 and January 1989, another five systems engineers and one computer technician (promoted from the shopfloor) were appointed to do the routine work on various parts of the CIM system, ranging from office computers and POMS through to modifying plc controls on the machines. At the same time, fearful of losing existing staff in a highly competitive external labour market, the existing systems engineers were promoted to a new grade of 'senior' systems engineer and given a hefty pay increase.

### The limits of full automation

POMS had always been regarded within the project team as the heart of the Aberdare CIM system (see Chapter 4). However, despite the extra systems staff, numerous simulated trials of the SEIAF software in the first half of 1989, and a few actual trials on the shopfloor, it gradually became clear to Wolstenholme, Peters and Dawson, and through them to Regan, that POMS was simply not going to be able to do what had been originally intended.

The detailed technical reasons for this have been discussed elsewhere (Clark, 1994). But, with the benefit of hindsight, two main problems can be identified. First, the idea of computer systems planning and directing every phase of the building wires business with only residual human intervention was almost certainly unrealizable with the level of technology and systems knowledge available at the time it was attempted. Indeed, with around 600 cable types, of which some 200 were fairly common, manufacturing building wires is not a mass production but a batch production industry, requiring regular changes of product type and materials. Every change of product requires changes in wire size, machine set-ups, dies and insulation colour, all of which inevitably increases the risk of faults or failures. Programming software from scratch to set parameters to produce every type of cable, select every mix of materials, and identify all possible faults and failures was, as one system engineer said, 'a nightmare'.

However, even had the idea been realizable, to go straight for full computer control was an extremely high-risk implementation strategy. For such an approach to have been successful it would have required not only a fully-trained workforce and management, but machines, machine controls and software which all functioned reliably and consistently immediately following installation and commissioning. At Aberdare, many of the machines, the machine control systems, the AGV transport system, and the POMS software were prototypes in the cable-making process. The likelihood that some parts would go wrong or require modification was bound to be high.

### The reasons for the decision to modify POMS

It took until the summer of 1989 for Regan, Dawson, Wolstenholme and Peters to admit to themselves that their original dream was not going to be realizable in the short term. It was then that they took the brave decision to draw back from full automation. The reasons for the problems with POMS were manifold. There were major communication problems with SEIAF personnel, not only as a result of language difficulties, but because of the geographical distance between Aberdare and Genoa. Of course, this would not have been a problem if the system design had been specified fully and accurately, but it was not. The SEIAF system designers were not experts in cable-making, and the Pirelli implementation team members were not experts in software design. Both sides reinforced each other in 1986 and 1987 in believing that everything was possible.

There is little doubt that Regan's often professed belief in the ideal of a fully-automated factory ('I believed in magic') was one of the big ideas which actually helped pull the factory and its staff through their most difficult period and out the other side. But in 1989, in many ways the worst year of all, it looked like things were never going to go right. Machine breakdowns were a daily occurrence and maintenance staff were spending most of their time patching up faults to maintain some semblance of production. At the same time, Wolstenholme, Dawson and Peters put in enormous hours to try and make POMS work. But the fact that none of them had direct access to software specialists (apart from SEIAF employees) who had the expertise to say whether what they wanted could actually be done, and if so how long it might take, meant that, with the benefit of hindsight, they almost certainly persisted with their belief in magic longer than they should have done.

## Full POMS becomes virtual POMS

By the summer of 1989, around the time of Martinez's arrival, they had become convinced that they would have to step back from full automation – at least temporarily – and devise an interim manual system of production which would enable them to achieve something like the required production targets. They decided on the following fairly conventional scheme. Scheduling would be done on paper by the production coordinator. He would generate a weekly production plan in collaboration with the commercial manager based on orders from the regional depots. Every morning he would draw up a 24-hour production plan together with the day shift manager, and this would generate a number of manufacturing orders (MOs) for particular types and length of cable product. The paper-based MO would be attached to the drum or bobbin on its way through the production process and signed off manually by the machine operators when they had completed their task and carried out the appropriate quality checks. AGV missions would be raised by operators manually, and POMS would draw on the MMS to tell the machine operators what materials were required. Hardly the brave new world of computer-integrated manufacturing!

In the meantime, Regan and his colleagues decided to go back to first principles and asked SEIAF to redesign POMS to achieve a more limited set of objectives. The new system was given the name v-POMS or 'virtual POMS', obviously an exaggeration of its closeness to the original concept (full POMS), but meant to signify Regan's and Dawson's continued commitment to the ideal. V-POMS was to have three main functions, to be implemented in successive stages in 1990: computerized production monitoring (providing information on what was being produced and how far each order had progressed); computerized inventory management (the monitoring and control of raw materials stock, work-in-progress, and finished goods stock); and computerized machine set-up data (for different product types and specifications).

Not surprisingly SEIAF argued that the changes that PG was now demanding represented a substantial variation from the original contract and that a new financial arrangement needed to be agreed. As we saw in Chapter 4, plant management systems (the equivalent of POMS) are the most risky and expensive parts of CIM systems, and at this time no other company in the world had designed such a system for a cable-making facility. Given the difficulties in specifying how long the design would take, being effectively tied to SEIAF, and being firmly committed to some form of POMS, Regan and Mancini decided that they had no alternative but to agree a new arrangement which increased the software/systems costs significantly before the contract with SEIAF was terminated in 1991.

## *Unforeseen consequences of the retreat from full automation*

The retreat from full POMS was a bitter pill for Regan, Dawson, Wolstenholme and Peters to swallow. But at least they had saved something from it. Neither Cole nor Martinez had been convinced that it was worth keeping at all. However, neither had any knowledge or experience of manufacturing software and they were prepared to accept, albeit reluctantly, the retreat to virtual-POMS.

However, the decision to accept a more manual system of production planning and sequencing had a number of unforeseen consequences. It was no secret that some of the production and planning managers, and many of the production operators, had never been fully committed to computerized management of production, partly through a reluctance to break with traditional practices and partly through a fear that it would not work (but not, according to interviews conducted in 1990, because of any widespread fear of loss of skills). The failure of full POMS seemed to confirm their doubts, and they were delighted to return to more manual systems of production. Many welcomed the switch to the more traditional focus on 'getting tonnage out of the door'. They were also becoming increasingly worried about the future viability of the factory (and their jobs) while it continued to make a loss.

Much of the resentment about POMS was directed at the systems engineers. They were seen by many producers as overpaid graduates who were distanced from the realities of shopfloor production, played about with computers, took long lunch hours, and every so often interrupted production with POMS trials that were doomed to failure. In contrast, for many systems engineers, the failure of POMS resulted not only from an overambitious and misguided original specification, but also from a lack of commitment and support among production staff, particularly some production and planning managers. Most believed that PG and the senior management team at Aberdare were suffering from a lack of nerve in their failure to stick with the original idea and see it through. Many saw the retreat from full POMS as the end of the experiment in computer-integrated manufacturing and a permanent return to more traditional methods of production. In 1990, four systems engineers resigned, and the remainder showed various degrees of frustration at what they saw as a marginalization of their role.

### Trials for v-POMS

Trials for v-POMS went ahead in the summer of 1990. They initially served to confirm both sets of prejudices. For many production operatives and production managers, POMS was simply an expensive hindrance to output. In the course of attempts to reintroduce automatic control of the AGVs, one producer had to wait up to five hours for an AGV to arrive to take away a bobbin (no manual intervention allowed!), and there was much cursing at the lost production. In contrast, systems engineers felt they were not being given enough time to see the trials through, and were having to carry the can for the system designers. They longed to be able to take over the system from SEIAF and modify it themselves. When interviewed in the summer of 1990, over 80 per cent of production operatives wanted POMS scrapped in its entirety. By the summer of 1992, there was almost unanimous support for it and praise for the way it had developed since 1990. The story behind this change will be told in Chapter 10.

# Changing Business Objectives and Performance Targets

1990 was the first full year in which the new factory reached anything like normal and stable patterns of working. As the end of the year approached, attention turned increasingly to how the factory had performed in the year, how the MD and sector managers would assess its performance, and what targets could be realistically achieved in 1991.

The overall objectives for the unit remained as they had been set by the PG board of directors in 1985: to maximize the return on capital employed by becoming the lowest-cost producer and by increasing market share. The main means of achieving this was investment in the most up-to-date CIM technology. Over and above this, sector managers wanted the factory to be a model for the rest of the group world-wide. This involved substantial experimentation and risk-taking, in particular investment in high levels of automation which was almost certainly not commercially justifiable in the short term. However, it was hoped the company and the group would achieve a pay back in the medium to long term in the form of a competitive advantage deriving from highly efficient, reliable, flexible, low-cost production. Even if some of the experiments proved to be failures, then the group as a whole would be able to learn from them.

By the beginning of 1990 Regan and Dawson had yet again failed to reach the immediate target of completely taking over production from the Southampton factory, which continued in operation, albeit at a reduced level, until December 1990. Although they had taken a number of measures in the first half of 1989 to try and achieve the original ideal of an automated factory, the advent of Felipe Martinez as the new MD for Pirelli's UK cable operations signified a shift in emphasis, both in the overall objectives of the Unit and in its immediate short-term objectives. Martinez had no particular historical or personal commitment to the experimental function of the site; indeed, he was convinced that if the factory was to continue at all, the first priority was to start producing substantial and regular output. In some senses, this shift in emphasis coincided with Regan's own view that the factory would have to 'deliver the goods' in the short term if they were to convince Milan that the experiment was worth pursuing in the medium term.

This did not mean that other objectives – apart from production output – were not important. However, given the special circumstances in which Martinez found the Aberdare factory in mid-1989, he left Regan in no doubt that the achievement of production and productivity targets in 1990 was the crucial precondition for the achievement of all other targets. In turn, Regan and his senior management colleagues made this abundantly clear at every business review meeting they had with staff (on these meetings, see Chapter 8). In October 1989, for example, commercial manager Peter McBride complained that sales were still limited 'due to production supply problems', and in July 1990 Regan reported that they were having difficulties with one of the site's major customers due to 'problems of unavailability of supply'. At the same meeting he also sounded a note of caution about the UK building wires market. He pointed out that, while the total market was in slight decline due to the recession, cheap imports from Eastern Europe and elsewhere were rising fast, and new UK cable companies were providing strong competition. So, although Pirelli's market share was rising well above the reduced share of 1989, they were not able to sell cable at the originally planned levels. As a result, the management plan sales target for 1990, expressed in terms of income from sales, was unlikely to be met. Nevertheless, everyone on site was left in no doubt that the needs of production were to be paramount. Again and again Regan stressed that it was absolutely vital to achieve the 1990 management plan target of 17,000 tonnes if the factory was to be seen in South Hampshire and Milan as a viable proposition for the future.

This 'heads down, bums up' approach, as one manager called it, was understandable to most staff. For many systems engineers, though, it was dispiriting. Pirelli's traditional organizational culture had always been extremely production-oriented, and many had hoped that Aberdare, with its total systems approach, would be different. In July 1990, one of them commented ruefully, 'tonnage is the one and only discipline in the factory'. This temporary suppression of many of the wider objectives of the Aberdare experiment was not to Regan's liking. However, most of the staff were unaware of the intense pressure under which he and his senior management team were being put from mid-1989 to the end of 1990. Regan had staked his reputation and his job on an assurance to Martinez in early 1990 that the production target for the year would be met in full.

## Christmas Eve 1990

After a relatively poor August and September 1990, the next two months showed excellent production figures. In October output at over 2000 tonnes was the best ever, as was the number of kilograms of cable produced per total employee hour – one of Milan's key internal productivity measures. These figures were particularly satisfying for Martin Dawson, as this was to be his last month as operations manager at Aberdare. He had learnt in the summer that he was being recalled to South Hampshire and promoted to the position of works manager of the telecommunication cables division, working as second in command to his old friend and colleague Stephen Cole. Dawson had given an enormous amount of time and energy to the factory over the previous four-and-a-half years. He had uprooted his family to move to South Wales, and was extremely reluctant to leave just as things looked like they were finally coming right. But, as he pointed out ruefully at the time, if you want to get on you do not turn down promotion opportunities offered to you by the managing director.

But what about the management plan output target of 17,000 tonnes? Early in the morning of Monday December 24, the day after Regan's birthday and the final working day of the year before the Christmas shutdown, the magic figure was reached. A great cheer went up! The final figure for the year was 17,058 tonnes. They'd done it. They had proved over the previous three months that they could consistently hit the required weekly tonnage. With continued machine improvements and more experience, they were sure they could raise output by a further 20 per cent as required under the 1991 management plan. Everyone went away for Christmas convinced that they had turned the corner.

Regan and his operations manager, Martin Dawson, had been in an extraordinarily difficult position in 1989 and 1990. They were under constant and intense personal pressure (from which they shielded the staff at Aberdare almost totally) to show positive results to justify the investment in new technology. In mid-1989 and early 1990, in particular, Regan was told at the highest level that his job was on the line. It needed personal courage – and the continued belief in many of the fundamental ideas and philosophies behind the Aberdare approach – to sustain him during what were very dark days. Perhaps the most sustaining feature of all was what Regan and Dawson felt to be – or, in Regan's comment on the manuscript in April 1994, 'knew to be' – the general success of their new human resource policies.

## Summary

In this chapter we have followed the first three years in the implementation of advanced technical change at the Aberdare factory. The original conception of a computer-integrated factory with a flat management structure presupposed that the new machine and software systems would run smoothly after installation and commissioning with little need for human intervention. In the event, there were substantial problems with the machine systems, and the modifications that were made to them intensified existing problems with the design and implementation of POMS, the plant operations management system.

In early 1989, a troubleshooter was sent from PG headquarters to help sort out the machine and production management problems. However, it was the appointment of a new PG managing director, with first-hand experience of managing the implementation of a new but more conventional building wires factory in Spain, that symbolized a significant change in the emphasis of the project.

This entailed playing down the experimental function of the factory and concentrating on the achievement of quantified production output targets. As part of this reorientation, and for many other reasons too, Regan and his senior management team decided to retreat from full automation and to insert a substantial degree of manual intervention in the production process. This sharpened the existing polarization between production and systems engineering staff on site, but also led to the achievement, on 24 December 1990, of the annual production output target for the first time in the history of the plant.

---

## Questions for Discussion

1 To what extent does the hierarchical structure of a typical CIM system imply a 'top-down' approach to CIM implementation?

2 Examine the key reasons for the retreat from full automation in 1989.

3 Discuss the relative significance of the various mechanisms and policies used to bring about a modification of the CIM system at Aberdare in 1989.

4 What criteria would you use to judge whether the modifications made were successful or not?

5 Compare and contrast the main characteristics of Pirelli's Aberdare and Spanish building wires factories.

# 6
# Designing a New Human Resource Strategy

## Thematic Focus: Human Resource Management – Models and Realities

Over the past decade or so, many claims and counter-claims have been made about the phenomenon of human resource management. Does it signify a fundamental break with the traditional way of managing people, with a greater emphasis on generating commitment, empowering individuals, and developing their skills and careers? Or are many of the new arrangements which go under the name of HRM – such as flexibility, team-working and performance-related pay – in reality less comprehensive, less empowering and more contradictory than the theory would suggest? Or is HRM even, as one commentator has suggested (Fowler, 1987), simply the discovery of personnel management by chief executives?

There are many problems with trying to assess such conflicting claims (see Clark, 1993b). Often, an ideal or rhetoric of human resource management is compared, not with the practice of HRM, but the practice of traditional personnel management (see Legge, 1989). In addition, there are not only many different styles of personnel management (see Purcell and Sisson, 1983), but many different versions of HRM: hard HRM, soft HRM, weak HRM, strong HRM (see Storey, 1992; Sisson, 1994b). In order to avoid a theological debate about the essence of HRM, its main features can be best understood by looking at the two basic forms of HRM practice currently in operation in the USA and the UK, and comparing them with traditional personnel management practice in these countries (see Keenoy, 1990).

*Traditional HRM* is exemplified by the long-standing practice of companies such as IBM, Hewlett-Packard and Marks & Spencer. Their approach to the management of people is based on the assumption that if an organization invests in individuals – for example, through sophisticated recruitment and selection techniques, systematic training and career development programmes – and pays and treats them well, they will respond with loyalty and a high commitment to work. As a spin-off, they will also be unlikely to want to join a trade union. These organizations have always adopted a proactive approach to human resource issues, and the specialist personnel or human resource function has always been represented in the boardroom and played a major role in shaping overall corporate strategy and organization culture. The objective of human resource policies in such companies is, to quote Keenoy, 'to ensure a stable, satisfied, and company-oriented labour force as a foundation for the cost-effective pursuit of market share and profit' (1990: 4).

Such an approach to the management of people is based on values which are both individualistic and unitarist: individualistic, in that they concentrate on the link between the individual employee and the organization and are generally hostile to trade union representation; and unitarist, in that the company is seen as the undivided focus of loyalty with no fundamental conflicts of interest between the company and its employees, between management and workforce.

Traditional HRM is thus quite distinct from the policy and practice of *traditional pluralist personnel management* as practised in the majority of medium and large UK companies in the post-1945 era. In this tradition, the organization's main task in the management of employees is not to seek their commitment, but their compliance in what is generally seen – at least as far as manual workers are concerned – as a low-trust, largely adversarial relationship. Traditional personnel management is also based on the idea of mutuality – the mutual recognition of plural interests in the enterprise and the need for mechanisms to reconcile these divergent interests in a peaceable way. The main method of achieving this reconciliation in the UK and USA has been through collective bargaining with trade unions – often representing different occupational groups – which are seen as key actors in the management of conflict within the enterprise.

The second form of HRM is what might be called *neopluralist HRM*. It represents a compromise between the personnel and human resource management traditions. The American social scientist Walton has described it thus:

> [It is a] strategy for eliciting commitment based on policies of mutuality . . . The legitimate claims of multiple stakeholders – employers, customers, and the public, as well as owners – are usually acknowledged. The fulfilment of many employee needs is taken as a goal rather than merely as a means to other ends . . . The common thread of the policies of mutuality is first to elicit employee commitment and then to expect effectiveness and efficiency to follow as second-order consequences. (cited in Keenoy, 1990: 4)

According to Walton, such strategies were first developed in the USA in the early 1970s, first in greenfield, mainly non-union sites, then subsequently in unionized manufacturing companies. They often involve a commitment to high levels of employee participation in the design of jobs and the management of quality. Where trade unions are recognized, relations with management tend to be cooperative and non-adversarial.

Since the mid-1980s, this type of neopluralist HRM approach has been increasingly evident in the UK in the rhetoric, if not always in the practice, of mainstream companies such as Ford, ICI and Lucas (see Storey, 1992). While the institutions of collective bargaining and collective representation have been maintained, their significance and power have been reduced. At the same time a number of more unitarist practices – more direct communication with the workforce, performance-related pay, quality improvement groups – have been introduced. Such companies exhibit what Storey has called the 'co-existence of two traditions' of beliefs and assumptions, an older personnel/industrial relations tradition, and a newer HRM one (1992: 30). So while 'neopluralist' HRM shares many features with 'traditional' HRM, it differs in three important respects: it tends to be pragmatic in its attitude towards collective bargaining and trade unions; it recognizes the legitimacy of divergent interests within the enterprise; and it opens up real possibilities for employees to exercise discretion and involvement, whether through joint decision-making, joint consultation, or simply access to information about the enterprise and its performance. While there are many different varieties of HRM in practice, most are permutations on these two basic forms. All stand in explicit or implicit contrast with traditional personnel management policies and practices.

On a slightly different note, some commentators have argued that the essential difference between 'personnel' and 'HRM' resides in the extent to which organizations take a *strategic* view of human resource or personnel issues (see Keenoy, 1990: 5–6). However, as Guest has pointed out, it is not that traditional personnel management has failed in the past to take a strategic view. What makes HRM distinctive is the particular direction and objectives behind its strategy.

Guest has suggested that the distinctiveness of HRM can be encapsulated in one basic proposition: that there is an organizational payoff for companies if they develop a combination of policies designed to meet four key human resource objectives: organizational integration, high commitment, employee flexibility, and high quality (1989: 42). Organizational integration implies a planned and coherent HRM strategy which is 'owned' by senior management and adopted by line managers in their operational decisions and interactions with staff. High employee commitment has two features: an attitudinal commitment reflected in a strong identification with the company, and a behavioural commitment to pursue agreed goals in day-to-day work activities. Employee flexibility means task flexibility and the breakdown of traditional job demarcations – the two main features of what is often called 'functional flexibility' – together with the use of a greater variety of contractual relationships with staff (full-time, part-time, temporary, sub-contracted services, and so on). Flexibility also requires an organization structure capable of managing innovation and adapting to continuous change. Finally, high quality refers not just to the quality of work and working life (working environment, job content, job satisfaction), but also to the quality of the workforce. This is reflected in high investment in skills and training, and in the quality of products and services provided (see Clark: 1993b, for further discussion).

If Guest's proposition about the organizational payoff of HRM is to prove valid, then specific HRM policies (for example, selecting employees on clearly-defined criteria using appropriate selection techniques) should enable organizations to achieve particular HRM goals – such as high commitment to work and functional flexibility – and particular organizational outcomes – such as low labour turnover, greater allegiance to the organization, and ultimately, improved performance and competitiveness. In the following chapters we will be tracing how Pirelli Cables went about designing new HR policies for its new plant at Aberdare, and assessing whether it achieved an organizational payoff from them in practice.

Before we do this, it is important to look in a little more detail at the take-up of HRM initiatives and the organizational reality underlying the influential rhetoric. Large-scale surveys and case study evidence in the UK suggest that, since the mid-1980s, the take-up of individual HRM practices – such as performance-related pay, team briefings, financial participation through profit-sharing or share-ownership schemes – has been significant (see Millward et al., 1992; Storey, 1992). However, many of these have more to do with the reassertion of management control and the decline in union power than with any positive commitment to human resource management (see Sisson, 1994b: 24–6). In addition, there is growing evidence that in a number of companies HRM policies such as merit pay and team briefing have fallen into disuse. Only in a small minority of cases is there any indication that these policies are 'integrated' into wider business or corporate strategies, as HRM theory would suggest. Finally, with the exception of companies such as Marks & Spencer and IBM, HRM practices are extremely rare in the majority of non-union companies, where unconstrained managerial prerogative reigns supreme and insecurity and lack of involvement are the norm for the majority of employees (see Millward et al., 1992: 363–5).

These findings are sometimes explained by reference to inconsistencies between the various elements of HRM, for example between team-working and individual performance-related pay. Others point to the fact that in many companies HR policies are largely subordinated to finance-driven business strategies. Yet others argue that the contradictions within HRM are inevitable due to the structural antagonisms between managers and employees inherent in the employment relationship (see the collection by Blyton and Turnbull, 1992, for examples of all these explanations).

There are few reliable longitudinal studies of how new HRM strategies are decided and implemented, how they change over time, and how deep-rooted and lasting are the changes they bring about. These are the core themes of the next three chapters. However, whatever form the changes take, there is general agreement in the research literature that the quality of management and managerial practice is crucial. Perhaps the most distinctive feature of HRM, and the one which marks it out from traditional personnel management's almost exclusive concentration on non-management staff, is its emphasis on managers themselves – their recruitment and selection, their performance targets, their motivation, training and development (see Clark, 1993b). As Guest has argued:

HRM places a premium on the competence of management, and if the changes associated with the 'new industrial relations' are to persist, then it will be partly because HRM policies provide the basis for management practice. At the same time pursuit of HRM is not risk-free; but to organizations driven by market pressures to seek improved quality, greater flexibility and constant innovation, HRM may appear to be an attractive option. (1989: 45)

This is an accurate description of the basic position adopted by Pirelli Cables as it set out to design a human resource strategy for its new Aberdare plant.

---

# Traditional Personnel Policies and Practices

Detailed consideration of the place of human resource policies in the overall strategy for the Aberdare site did not begin in earnest until after the final decision to go ahead with the project in September 1985 (see Chapter 2). By then, a number of assumptions had been made about various aspects of the new HR policies. Both the Roberts report of June 1984 and the recommendations to the PG board of directors in June 1985 had assumed a major reduction in staff numbers, to 150 and 170 respectively compared with over 300 required to produce building wires at the Southampton factory. Both had also assumed that there would be a need for flexible or group working, and a tighter and leaner management and organization structure. In addition, the possibility of achieving a single-union agreement had been an important element in the decision to locate at Aberdare (see Chapter 3). All these assumptions and initiatives had been taken without any involvement of the PG personnel department. Before we look at the growing involvement of personnel specialists, it is worth tracing briefly the evolution of the personnel function in Pirelli General, as this was the context from which specialist personnel managers were able to exert their influence.

## *Creating a specialist personnel function*

Until 1960 there was no separate personnel department in the company. Payroll and personnel records, manually recorded and stored, were the responsibility of the corporate accounts department. Each factory had a senior clerk or 'labour officer' – answerable to the accounts function – who had some involvement in recruitment and discipline and kept basic personnel records. However, with the increasing power of shopfloor unions and the growth of white-collar unionism, PG corporate managers (but not managers in the operating divisions, who resented the loss of influence) decided in the late 1950s that it was time to institute a professional company-wide personnel department to deal with these matters. The first two appointees took up their positions in 1960.

The first company personnel manager was Richard Jones. He had grown up in the South Wales valleys and worked as a personnel manager at the Hoover factory in Merthyr Tydfil, 10 km from Aberdare, before applying for the job at PG. His assistant, Peter Barnett, had previously worked for GEC. Over the following decade Jones and Barnett were responsible for establishing the basic personnel framework which lasted, with one or two notable exceptions, until the mid-1980s. It had the classic features of what Purcell and Sisson (1983: 115) have called the 'constitutionalist' style of traditional pluralist personnel management: a commitment to mutuality; employee relations policies centred on the need for stability, control and the institutionalization of conflict; a substantial role for unions; and the defence of management prerogatives through highly specific collective agreements and careful attention paid to their administration on the shopfloor (see Sisson, 1989: 9–11).

### Evolving separate industrial relations systems

During the 1950s and 1960s, separate systems of industrial relations evolved for the three main groups of non-management employees in the company: production workers, maintenance workers, and white-collar staff. Production workers had their basic terms and conditions of employment determined by a national Joint Industrial Council (JIC) which covered most of the UK cable industry. The JIC was composed of members of the national employers' association and full-time officers from the recognized trade unions, of which the two largest were the Transport and General Workers Union (TGWU) and the General and Municipal Workers Union (GMWU, later GMB). From 1969, production workers were also covered by a national job evaluation scheme which divided them into one of eight separate job grades.

Richard Jones played a leading role as an employer representative in the national JIC negotiations. He was also responsible for negotiations at company level over supplementary payments and bonus schemes for production workers, and he chaired company-level consultative meetings with the unions on questions such as annual closures and facilities for shop stewards. By the early 1980s, the TGWU was the dominant union for production workers in PG, with the exception of the factory at Aberdare, where the existing recognition agreement with the GMWU had been allowed to continue after PG had acquired the site in 1971.

There were no parallel national negotiations for maintenance workers in the cable industry, and Jones took the lead in company-level negotiations with their trade unions. There were two unions representing the two main craft groups, the engineering union (AEU, later AUEW, now AEEU) for mechanical maintenance, and the electricians' union (ETU, later EETPU, now AEEU) for electrical maintenance; the TGWU represented semi-skilled maintenance employees. In the early 1970s the company concluded a closed shop agreement with all the shopfloor unions, so making union membership a condition of employment for all manual workers. The shopfloor was thus a totally unionized environment, with different unions representing the three basic occupational groups (production, mechanical maintenance and electrical maintenance), who were employed in a multiplicity of jobs and grades with strict demarcations within and between them.

While Jones dealt with shopfloor industrial relations, Barnett took responsibility for 'staff' employees, which included everyone from basic clerical grades and technicians through to supervisors and middle and senior non-executive managers (the terms and conditions of the small number of executive managers were handled jointly by Jones and the PG managing director). From the late 1960s, following union pressure, PG recognized separate unions for the three main categories of white-collar staff: clerical and secretarial (represented by ACTS, the white collar section of the TGWU), technicians (TASS, now MSF), and supervisors and junior managers (ASTMS, now also MSF). This left a number of middle and senior managers in a residual category of 'senior monthly staff' or SMS. Eight SMS grades were created, and their jobs were assimilated onto the Hay evaluation scheme,[1] which was already in operation for executive staff.

## Expanding the specialist personnel function

As plant level industrial relations grew in complexity in the 1960s and 1970s, the personnel function within Pirelli General expanded significantly. In the mid-1960s, following the statutory imposition of a training levy on employers in the UK engineering industry, a new post of corporate training manager was created to coordinate the company's initiatives on training. A number of specialist posts were also created to deal with the training of particular occupational groups. By the mid-1970s, too, company-level negotiations with production and maintenance workers were becoming so time-consuming that specialist responsibility was hived off from Jones's job into a new position of industrial relations officer. In 1981 this post was upgraded and renamed employee relations manager.

Parallel to these developments, the old plant 'labour officers' were upgraded to divisional personnel officers and gradually assumed responsibility for all day-to-day personnel matters affecting shopfloor workers in their factory. However, they still had virtually no role in dealing with white-collar or senior monthly staff in their divisions, who remained the responsibility of the corporate personnel department. By this time Peter Barnett was spending most of his time dealing with the expanding number of senior monthly staff in the company, and a new post of staff manager was

[1] The Hay evaluation scheme, run by the multinational company Hay-MSL, is a generalized method for producing an independent evaluation of managerial jobs. It is used in companies across the world. Jobs are awarded points on a standard range of criteria, and individual companies then devise what they believe to be the most suitable banding of grades to create a managerial pay and grading structure for their purposes. Hay also conducts regular national and international salary surveys of companies affiliated to its scheme world-wide, and these are used as a benchmark for companies to set their own management salary levels.

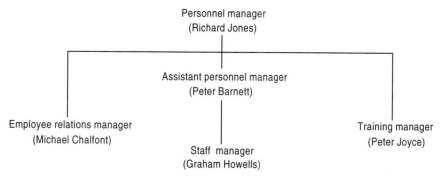

**Figure 6.1**    *The corporate personnel department of Pirelli General, 1985*

created to assist him in dealings with white collar unions and the associated personnel administration.

So, by mid-1985, when detailed discussions about the new human resource strategy for Aberdare were about to begin, the corporate personnel department consisted of a personnel manager dealing with executive staff and with national negotiations for production workers, an assistant personnel manager responsible for senior monthly staff, an employee relations manager responsible for company-level negotiations with production and maintenance workers, a training manager, and a staff manager assisting with all matters affecting white-collar staff, including company-wide negotiations with white-collar unions (see Figure 6.1).

The position of staff manager was occupied at this time by Graham Howells. He had joined Pirelli General in 1976 as a junior personnel officer in the corporate personnel department, having worked for the Eastern Electricity Board for two years after leaving university. In 1980 he became divisional personnel officer in PG's large power cables division, returning to the corporate personnel department in 1983 as staff manager. Jones, Barnett and Howells were the key corporate personnel specialists involved in helping design the framework for the Aberdare human resource strategy between 1985 and 1987.

## Signalling the beginnings of a new approach to personnel management

Two developments at PG in the early to mid-1980s already marked the beginning of a change in the company's approach to personnel management and industrial relations. First, under strong pressure from managers in the operating divisions, who wanted greater freedom to determine their own personnel policies, PG decided in the mid-1980s to withdraw from national industry-wide negotiations. This it did in 1987, together with BICC and Delta, its two main UK competitors. This spelt the end of national multi-employer bargaining in the cable industry. It also led to a strengthening of PG's own corporate personnel department, which now had responsibility for setting personnel and industrial relations policies for all grades and all levels of staff in the company. In part to recognize this change, the post of employee relations manager was upgraded in 1986 to assume responsibility for all company level relations with shopfloor and staff unions. In June 1986 Howells, now aged 33, took up this new position when Chalfont moved to a senior management position in one of the operating divisions.

The second new development was the signing in 1983 of PG's first comprehensive plant-level collective agreement. This covered the brand new submarine cable factory which had been built in 1982–3 to produce cross-channel power cables to interconnect the British and French national electricity supply systems (for details see Jary, 1990: ch. 7). The agreement was negotiated jointly with the three manual workers' trade unions over 10 meetings, and involved on the company side the three main operational managers – divisional manager, works manager and project manager – and three personnel specialists: Jones, Chalfont and Howells (the latter in his capacity as personnel officer for the power cables division). The agreement had a number of unique features for the company:

- It covered production workers and mechanical and electrical maintenance staff and was negotiated and signed jointly by their three unions.
- Its first section listed a number of general operational requirements, including 'full flexibility of personnel' across and within traditional demarcations (this was modified subsequently to a commitment that workers in different jobs would help each other out as and when required).
- There were only four – rather than the nationally agreed eight – production jobs, and three of the four were graded equally.
- There was one unified payment system for all manual workers.
- Skilled production and maintenance workers were paid the same basic wage.

However, the agreement also contained a number of more traditional features:

- It only covered manual workers.
- It still retained separate categories of maintenance staff.
- It remained multi-union, with separate unions for production workers and the two separate maintenance crafts.
- It included, in addition to basic pay and shift allowance, a whole range of extra bonuses and allowances, including a 'stable bonus' in return for flexibility, a lump-sum for completing the contract on time, and a callout allowance, as well as shift handover, meal break, cable-loading, and callout payments.
- It had a form of status quo procedure which required the company to consult in advance on changes to staffing levels, job structuring, and major

production and equipment improvements. If agreement could not be reached, the company agreed not to implement changes until the procedure had been exhausted.

- It confined itself to basic industrial relations issues, containing no reference to wider human resource philosophies or policies (such as training, skill acquisition, appraisal).

In this and many other respects, it represented a halfway house between the traditional pluralist personnel management approach and a new kind of HRM strategy.

## The Evolution of a New Human Resource Strategy

The HR strategy for the new Aberdare plant was developed over a period from the summer of 1985 to the spring of 1987. It was dominated by two guiding principles. First, it took as its starting point the operational requirement for computer integration of all aspects of the business and full flexibility and adaptability in the production process. Any innovation by personnel specialists would have to conform to this overarching 'technology-driven strategy' and provide a clear payoff in terms of the achievement of overall business objectives. Where operational requirements did not 'dictate' specific personnel policies, however, then the second guiding principle would come into play, the requirement – laid down in 1984 by Arthur Stokes (see Chapter 1) – to review all existing practices and to be experimental and innovative where appropriate. To adopt Karen Legge's well-known distinction (1978: 79–85), personnel specialists were expected to be predominantly 'conformist innovators', but they did have discretion to act as 'deviant innovators' where operational and technical requirements allowed.

Richard Jones was a key figure in the design of the HR strategy. Born and brought up in South Wales and nearing the end of a long career with Pirelli, the Aberdare project gave him a new lease of life. At a time when PG was about to withdraw from national collective bargaining – the focus of much of his work over the previous two decades – the project was not only challenging, but also represented a return to his Welsh roots. He had two initial inputs. First, in April 1985 he commissioned a background paper on the various personnel management options for the new factory from Stanley Russell and Alistair Tate, director and deputy director of the Work Research Unit.[2] The 29-page paper – produced by Russell and Tate

---

[2] The Work Research Unit was a small group of around eight government civil servants, mostly with experience of industry and commerce, whose job was to publish background papers and shorter bulletins, hold conferences, and give advice to management and unions on issues such as job content, work organization, communications, consultation, payment systems, and the quality of working life more generally. At the time Jones approached them, they were in the process of transferring from the Department of Employment to the Advisory, Conciliation and Arbitration Service (ACAS), where they became a unit within that organization's more general advisory services. The WRU was dissolved in 1993 following government spending cuts and ACAS's increased workload in the area of individual conciliation.

following an extended meeting with David Regan about the technical and operational requirements for the new plant – was sent to Jones in July 1985, and led to a series of further meetings with Russell, Tate and other ACAS officials over the following 18 months. Many of the options set out in the paper in areas such as management style, job structures, work organization, payment systems, employee involvement, and union recognition, found their way into the Aberdare human resource strategy, as we shall see.

### Deciding on the functional composition of the senior management team

Jones's second early input related to the senior management structure of the new site. David Regan's most pressing priority following the approval of the project by the PG board in September 1985 had been to appoint his senior management team. All the other factories in the company had five direct reports to the divisional manager (Regan was divisional manager-designate of the new plant): production, sales, personnel, accounts, and quality assurance/technical. Production and sales were clearly crucial to the success of the new site and would need to be part of the senior team. Regan was also convinced that there should be a new senior post of systems manager because of the strategic importance of the new CIM system. However, he was totally against the appointment of a quality assurance manager. He believed that quality management should not be hived off into a separate department, but programmed into the computer-integrated manufacturing system, as well as being a fundamental responsibility of all employees. At the same time, he was not really convinced that personnel and accounts – which he saw as basically administrative functions – needed to be senior direct reports.

Despite his strongly held views, Regan decided to approach Richard Jones for advice. Jones was the longest-serving head of a corporate department at Pirelli General and had shown himself to be broadly sympathetic to the new venture. At the time he also enjoyed a high standing both within PG and at group level in Milan. Both Regan and Jones soon agreed that, given the tight staffing levels for the new plant which had been one of the preconditions for its acceptance by the board, there would need to be excellent communication links between all functions and levels and the number of direct reports to Regan should be kept to a minimum. However, neither had a clear view of how to achieve this nor which functions should actually be part of the senior management team. It was uncharted territory for both of them.

Eventually, they decided that since personnel would play a major role in setting up the factory – devising detailed personnel policies, dealing with the question of union recognition, recruiting and selecting new staff, and helping establish a new organization and work culture in the plant – they would create a new senior post of personnel and administration manager (later shortened simply to administration manager), to whom both the site accountant and purchasing manager would report. Once the factory was up and running, they expected that the personnel role would decline in importance and the site accountant would be likely to replace the personnel specialist in the senior management team (this is in fact what happened in 1989).

In late 1985 they went to Mancini with this proposal. Given Mancini's own commitment to reduce the number of direct reports, as well as his engineer's scepticism of the role of accountants, he agreed. Accounts manager George Sherfield was less supportive of the decision, but as it had Mancini's blessing and he was certainly not in good favour following the board meetings of June and September 1985 (see Chapter 2), he did not feel able to oppose it.

Mancini refused, however, to agree to Regan's proposal for a 'systems manager' to be part of the senior management team. He argued that this would lead to problems of overlap and coordination between the manufacturing and systems departments in what was supposed to be an integrated factory (the 'polarization of expertise' problem discussed in Chapters 4 and 5). It was therefore decided to create a four-person senior management team: Regan as divisional manager-designate; an operations manager covering production, maintenance and systems; a commercial manager covering sales and marketing, including the sales force of 50 attached to the then 10 regional depots; and an administration manager covering personnel, health and safety, accounts and purchasing. The small production planning section of Aberdare would have a dual reporting system, to both the operations and commercial managers, in order to avoid polarization between production and sales.

### Selecting and training the senior management team

In late 1985 Regan and management staff personnel specialist Peter Barnett set about drawing up job descriptions for the new posts. All three jobs, and particularly that of operations manager, appeared at first sight to be at least on a par with the comparable positions in the other factories, all of which were graded at executive status. However, there was a large degree of uncertainty about the nature of the new jobs and the total workforce was expected to be substantially smaller than in the other divisions. For these reasons, none of the three senior posts was initially graded at executive level, although Regan and Barnett assumed that when the operations and commercial managers had settled in to their new jobs, they would be raised from the top of the senior management structure (Grade 6) to executive status (this is in fact what happened in 1987). In contrast, the administration manager was only graded at 5. However this did represent a clear recognition of the higher status of the personnel function at Aberdare, since the personnel officers in all the other operating divisions were graded at 3.

Jones and Barnett then set about identifying potential internal candidates for the jobs. The post of operations manager was offered to the front-running internal candidate, Martin Dawson. He was only 32 years old, but had substantial experience in the company. He had also worked previously with Regan at Sterling Greengate in the early 1980s, and was at the time production manager in the general wiring factory at Southampton. However, before he was prepared to sell his house and move his young family to Wales, he did have to be offered some inducements – such as a company car and other executive fringe benefits – as well as assistance with his house move. In fact, Regan, Barnett and Jones soon came to realize that, despite the obvious attractions of the new 'state-of-the-art' factory, the idea of moving from the affluent and green South of England to what was perceived as industrial, depressed and 'far-flung' Aberdare was not attractive to many potential candidates for the other two senior positions. Eventually, the commercial manager's job was offered to a young, bright, abrasive Scot, 29-year-old Peter McBride. At the time he was one of the existing regional sales managers for the general wiring division and the post offered a double promotion. He accepted with alacrity.

As for the post of administration manager, this was offered to three internal candidates – including Graham Howells – before it was eventually accepted by the personnel officer in the telecommunication cables division at Bishopstoke, David Lane. Lane had joined Pirelli just over a year before, having previously worked as company secretary in a City of London shipping firm. Lane had had to think long and hard before accepting the offer. He had just been selected as prospective Labour parliamentary candidate for the Eastleigh constituency – in which most of the PG factories were located – and had hopes of a future political career. However, after talking to David Regan and discovering a kindred spirit, he accepted. He was very much like Regan, personable, imaginative and charismatic. His strength lay in ideas rather than detail. He was by no stretch of the imagination a traditional, conservative, play-it-by-the-book 'Pirelli man'. It was Lane who, over the next 18 months, together with Jones, Howells, Regan (and to a lesser extent Barnett), constituted the informal personnel working group which put the flesh on the bones of the new human resource strategy.

1 April 1986 was an important date in the overall development of the project. On this day David Regan took formal charge of the Aberdare site, the old factory was razed to the ground by bulldozers, and Dawson, McBride and Lane began their new jobs. However, only Lane was able to join Regan (the two self-declared 'crumblies' of the senior management team, aged 48 and 47 respectively, compared with Dawson at 32 and McBride at 29) to concentrate totally on his new post. The old Southampton factory had to be kept in production until the Aberdare plant became operational, and Dawson, who was production manager there until 31 March 1986, still spent some time each week checking on production performance. Meanwhile, commercial manager McBride spent most of 1986 and part of 1987 learning his new job, either working in the divisional sales office in Southampton or travelling around the country visiting customers and staff at the various distribution depots (see Figure 3.1). Indeed, it was not until 1988 that McBride started to build up the commercial department in Aberdare.

In the spring of 1986 Barnett worked out a personal training and development programme for all four senior managers. Dawson attended short external courses on various aspects of CIM systems, and a two-week course at Sundridge Park Management Centre on work management and new technologies. He also participated in a number of short courses on group working, employee involvement, action-centred leadership and team briefing run by the Industrial Society, and in-house courses on employment legislation, time management, and interviewing and assessment techniques. McBride and Lane also went on the same Industrial Society and in-house personnel management courses, and McBride attended a two-week Sundridge Park course on Marketing Management and Business Development.

### Excluding management staff from the new human resource system

Meanwhile, within a month of starting his new job, David Lane had prepared the first of what were to be many drafts of a paper entitled *Building Wires Business Unit – Personnel Policies*. It began with some general statements about the basic objectives of the new Unit and its underlying personnel philosophy. All were imbued with the importance of functional flexibility – task flexibility and breaking down traditional job demarcations – and the need to keep staffing levels to a minimum. Virtually all of the detailed proposals in the paper were geared to non-management staff, not only because there were going to be more of them, but because it was here that the greatest problems (and innovations) were envisaged. This concentration on non-management staff led, if not to a neglect of issues relating to management staff, then certainly to less intensive discussions in the personnel working group about the qualities the line managers would need and the detailed tasks they would have to carry out, particularly in the areas of team-working, management style and employee motivation.

As Lane's paper of 30 April 1986 made clear, the personnel subgroup decided at an early stage to treat managerial staff quite separately from non-management staff. There were a number of reasons for this. While the external labour market for recruitment of non-managerial staff would be exclusively confined to South Wales, the labour market for managerial staff would almost certainly be national, requiring nationally competitive terms and conditions of employment. In addition, while first-line managers had been included in union bargaining units across the company since the early 1970s, Jones and Barnett had become increasingly convinced of the need to take them out of collective bargaining and to break away from the customary practice of reserving such jobs for promotion from below. They also wanted to treat them as managerial rather than supervisory jobs and subject them to Hay job evaluation. The *tabula rasa* of the greenfield site gave them the opportunity to do this.

Barnett had also been trying over the previous decade to develop a long-term company-wide policy on management development and succession planning. This required corporate coordination and control to be effective. Finally, the Aberdare factory was intended to have only a small number of line management positions and levels. Unless the managers could be involved in a company-wide management development programme, they would have little chance for internal promotion and could be lost to the

| Pirelli General (1970) | Pirelli Aberdare (1986) |
|---|---|
| Divisional manager | Divisional manager |
| Works manager | |
| Production department manager | Operations manager |
| Assistant department manager | |
| Superintendent | |
| Senior foreman | Shift manager |
| Foreman | |
| Chargehand | |

**Figure 6.2** *Comparison of production management structures, 1970 and 1986*

company. Taken together, these arguments were convincing to the personnel specialists in the working group, and were accepted by Regan without demur. However, as we will see in the next chapter, the continuing corporate control over the job accountabilities and conditions of employment of senior monthly staff was to become a source of frequent friction between the Aberdare senior management team and the corporate personnel department.

## Designing the organization structure

In Lane's April 1986 paper, the only matters for debate about managerial staff were the specific job functions and the number of levels in the hierarchy. Here again there was broad agreement within the personnel working group that the numbers and levels should be kept to a minimum. In a typical early 1970s Pirelli General factory there had been eight levels of production management. By the late 1970s, following some restructuring, there were still six levels. In Lane's initial paper, and in all subsequent drafts, there were to be only three levels of production management at Aberdare, as computer integration was intended (as we saw in Chapter 4) to relieve them of many of their traditional tasks. A comparison of the typical PG production management structure in 1970, and the new Aberdare structure in 1986, is contained in Figure 6.2.

The number and type of middle and junior management positions at Aberdare went through many iterations. In the first draft of April 1986, Lane had suggested three positions which were subsequently deleted: quality assurance manager (deleted on Regan's insistence), office manager (Regan was highly anti-bureaucratic in outlook and argued that all managers should manage their own office work) and distribution manager (distribution was to be subcontracted to PG's wholly-owned subsidiary company, Meachers Transport). After a series of minor modifications over the following six months, the organization structure was agreed in November 1986. This operated until 1989 without alteration (see Figure 6.3).

In addition to these positions, there were to be nine specialist senior monthly staff posts: six systems engineers to work on CIM hardware and software, and three production engineers to assist the shift managers on

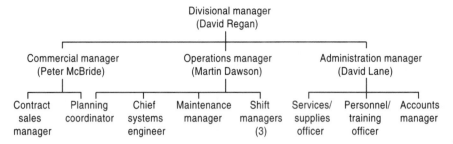

**Figure 6.3** *Line management structure at Aberdare, 1986–9*

production and development problems and to deputize for them in the case of illness or holidays. Lane and Barnett drew up detailed job descriptions for each post in consultation with the relevant senior manager, graded them under the Hay Evaluation Scheme, and began a search to fill the posts.

As with the senior management positions, it again proved difficult to attract internal candidates to move to Aberdare. As a result, the eventual appointees were rarely the first-choice candidates from within the company – as might have been expected, given its high prestige – but second, third or even fourth choices. The location of the site in Aberdare was clearly the decisive factor which hindered PG from attracting what it believed to be the best internal candidates for the job. This is in stark contrast with the experience of the Pirelli Group's new Spanish building wires factory, which was located just 2 km from other main factories and was therefore able to handpick those who were felt to be the best managers and staff for the new jobs (see Chapter 5).

### Selecting the line managers

Eventually around half the new management positions at Aberdare were filled from within PG, and half from outside. The latter included three Welshmen – the site accountant, the maintenance manager (recruited from Sony) and one of the shift managers. Only one of the managers in the new plant had worked in the old Aberdare factory – as a production superintendent – and Regan had in fact refused to appoint another of the old Aberdare managers when Jones had suggested him for a position. Regan clearly wanted a clean break from what he saw as the old-style, low-trust, heavily hierarchical management and industrial relations practices of the old site. In all, there were to be only 14 line managers at Aberdare, and just 3 shift managers for 60 production operatives. This substantial span of control, with no chargehands or team leaders, was clearly going to place a high premium on the self-reliance and autonomy of the operators (for the outcome see Chapter 7).

## Setting the framework for the new human resource strategy

In looking at the elaboration of the human resource strategy for the new site, two things are immediately striking. First, it deliberately involved collaboration between corporate and divisional personnel specialists and the unit's new divisional manager-designate, David Regan. This meant that while the strategy was strongly personnel-driven, it was also directly influenced and 'owned' by the senior general manager on site (see Yeandle and Clark, 1989a). Second, from the outset of detailed deliberations in April 1986, the personnel working group was committed to a shared set of business objectives and a shared underlying human resource philosophy. These are clearly reflected in the first draft of *Building Wires Business Unit – Personnel Policies*, prepared by David Lane after consultation with David Regan and Richard Jones. It was circulated to all members of the personnel working group on 30 April 1986:

### The Aberdare Human Resource Strategy: Background and Objectives (April 1986)

#### PERSONNEL POLICIES – BACKGROUND

The nature, location and circumstances of the building wires unit on the Aberdare site provide an opportunity to review all aspects of personnel policy and to adopt the most appropriate combination of policies and procedures. All of the unit's business systems will be computer integrated and the prime objective of the significant investment is to change the current loss-making position into a position of profit. Among other things, this means that the use of the human resource must be optimized. Therefore the company will require maximum flexibility from its employees within the unit. This flexibility will be best achieved by a combination of encouragement of team work and recognition of the employee as an individual. This is the underlying theme of all the proposals contained within this memorandum.

#### BUSINESS OBJECTIVES

To create a profitable and expanding business for the benefit of the company, employees and the community by maximizing efficiency and optimizing quality through establishing policies, systems and culture which elicit from employees commitment, conscientious endeavour, adaptability and initiative.

All the new personnel policies followed from these principles. This should not, however, lead us to underestimate the amount of work which needed to go into elaborating the detailed policies, many of which could not be simply read off from the wider business strategy. As we shall see, there was also ample scope for tension within and between various business, operational and personnel objectives.

## The involvement of the Work Research Unit

The ACAS Work Research Unit (WRU) played a major role in fleshing out the detail of many of the new personnel policies. Apart from preparing the 29-page background paper mentioned above, the WRU director and deputy director contacted a number of companies such as BICC, Black and Decker, Cummins Engines, Hitachi, Kodak and Nissan, which were developing new personnel initiatives in issues which seemed appropriate to Aberdare. They then arranged for various members of the personnel working group to visit these companies to talk to senior managers and learn from their experiences. All these companies (including one of PG's main competitors, BICC) were happy and willing to talk in some detail, even though Regan, Lane, Jones and Howells were open about their intention to adopt the best of the practices they found in order to improve their overall competitive performance. It would have been difficult to imagine a similar openness in the field of technical or product development. Clearly, human resource practices were still not seen as having the kind of direct payoff for competitive performance as new technologies or new products.

Three areas of personnel policy for the new site were strongly influenced by the WRU. First, there was the idea of integrated payment systems, on which a leading WRU member, David Grayson, had published a working paper in mid-1986 (Grayson, 1986). He argued that payment systems should be designed to integrate all members of staff around a common objective ('organizational integration') rather than, as with traditional bonus schemes, divide employees from one another according to output or occupational or departmental boundaries.

Linked to this was the idea of skill-based pay. Payment systems tell employees what an organization regards as its most important objectives: individual effort, team effort, output, quality, skill. If the prime objective was to encourage flexibility and initiative through continuous learning and training, then the payment system ought to be designed to encourage training and the acquisition of skills. In January 1986 Cummins Engines had initiated a new system of skill-based pay in its Darlington factory, and a visit to this factory organized by the WRU, together with reports on the experiment in the industrial relations press, played a major part in persuading the personnel working group to adopt this system for Aberdare. Finally, the WRU recommended in its July 1986 paper that if the company wanted to encourage employee commitment, it should adopt a range of employee involvement measures, including direct communication with the workforce and full disclosure of commercial and financial information.

## Developing the pay and grading structure

By June 1986, the broad outlines of the new personnel policy were clear, with the exception of two issues: union recognition (discussed in Chapter 7), and pay and grading. As to the latter, David Regan was initially in favour of having just one grade of non-management staff, all paid the same basic wage. He felt this would encourage staff to pull together and recognize their equal if different contribution to the work of the factory. The only differences would be in terms of shift allowance and skill-based supplements. However, the corporate personnel specialists, Jones and Howells, mindful of the importance of different labour markets for different grades of staff, were not convinced. To resolve the question it was agreed that Howells (now employee relations manager) and Lane (now spending more time in South Wales, where he had bought a small cottage) should make detailed inquiries about the 'going rate' for various types of jobs in the South Wales labour market and report back.

After gaining intelligence from a number of sources, including the Engineering Employers' Federation and two management consultants, a compromise was finally agreed which retained the idea of a basic grade for production, maintenance and office staff (Grade A), but instituted a second grade (Grade B) for a small group of employees where the going rate in South Wales was much higher and/or the job responsibility much wider. It was intended that this would include a handful of high grade technicians skilled in computer systems and electronics, and a few specialists in administrative areas such as home and export sales and accounts.

## The Human Resource Strategy Outlined

On 5 December 1986 Lane was able to send Regan, Dawson and McBride a detailed report on the new unit's proposed 'human resource' policy (the first time the term had been used in the deliberations), seeking confirmation that the policies should be adopted and implemented. Lane's document reaffirmed the centrality of employee flexibility, outlined a phased recruitment programme for 1987/8, and listed a number of specific policies:

- *Pay*: two non-management pay grades; shift allowance for staff working on shifts; skill-based pay for all based on the acquisition of skill modules; a 6-monthly performance bonus for all once the plant became profitable.
- *Hours of work*: 37.5 hours for all.
- *Holidays*: 25 days plus public holidays for all.
- *Sickness*: all employees covered for 4 months on full pay, from 4 to 6 months on half pay.
- *Pensions*: common arrangements, although senior monthly staff would also have life assurance as well.
- *Overtime*: for non-managers only, to exclude shift handover arrangements, meal break cover, and short periods of overtime to meet operational requirements.
- *Facilities*: common car parking, meals, presence recording and medical screening.
- *Training*: extensive off- and on-the-job training for all staff (see Chapter 7 for details).
- *Communications*: regular team-briefing, a 3-monthly general staff meeting, and a monthly business review meeting.

This was the core of the human resource policy which was eventually incorporated into a single-union agreement in May 1987 (see Chapter 8 for details of the agreement and the negotiations over union recognition and representation).

In concluding this chapter, there are two further human resource issues which arose at this time and were of particular interest in the context of the project as a whole. First, Lane's December 1986 paper contained a revised estimate of the staffing levels required for the new plant, now projected to be not 170 or 150 (as in 1985), but 115. Lane pointed out that this figure presupposed full flexibility of labour as soon as the plant was operational. He also stressed that it was totally dependent on the full automation of cable manufacture and the plant operations management system (POMS):

> The proposed complement reflects the significant reductions in manual inter-vention in the production process and in the level of clerical activity, compared with the current methods of working. In order for this complement to be sufficient it will be essential to achieve the planned levels of system support and process automation.

Second, it is notable that neither Milan nor the PG board of directors played any role in the evolution of the human resource strategy. As a matter of courtesy, Richard Jones had kept the group's central human resources department in Milan informed of progress, and had also sent them copies of major internal policy papers. But Jones was trusted as one of the most reliable, long-serving senior managers and, in any event, with the exception of the appointment of very senior managers, personnel policy was seen within the group as a matter for national company autonomy.

As for sector management and the UK board of directors, in December 1986 Stokes (still cable sector general manager, but about to retire and replace Lord Westgate as Pirelli General company chairman) asked Mancini for an update on two issues: the selection of the trade union for recognition and the proposed payment system. Following this request Mancini asked Jones – who asked Howells – to prepare background papers on the two issues. This was duly done, and the papers were sent to Stokes in February 1987. In a covering letter Mancini requested that, subject to any comments, decisions on these questions should be taken 'at management level within the company' and not submitted to the full board. In particular, he stressed that the process of selecting the union was well advanced and should be left to the corporate personnel department and operational managers at Aberdare 'acting on their best assessment of the South Wales industrial relations environment'. And this, as we shall see in Chapter 8, is what happened.

## Summary

This chapter has followed the process of designing a new human resource strategy for the Aberdare plant between 1985 and 1987. The first section looked at the personnel policies operating in Pirelli General's Southampton factories. This exemplified what Purcell and Sisson have called a 'constitutionalist' and pluralist style of personnel management with a strong role for PG's corporate personnel department in regulating industrial relations in the divisions. Two developments in the mid-1980s – the conclusion of a comprehensive collective agreement to cover all manual workers in the company's new submarine cable factory, and the decision to end national bargaining in the cable industry – signified the beginnings of a break with old-style personnel management and industrial relations in the company.

However, the new human resource strategy for the Aberdare plant represented a radical break with past practices, except in relation to management staff, who remained part of the company-wide personnel system. The strategy was developed by a small working group consisting of personnel specialists from the company and the division, together with the divisional manager-designate of the new site. They met at regular intervals, drew heavily on advice from the ACAS Work Research Unit, and visited a number of companies who were pioneering new developments in personnel policy at the time. The result of their extensive deliberations was a clear statement of personnel objectives and their relation to wider business and operational strategy, a lean and flat organization structure, and a comprehensive package of interrelated human resource policies. Unlike the technology strategy, there was little or no involvement from sector or group management in Milan.

## Questions for Discussion

1  How important was it for the evolution of the Aberdare human resource strategy that the plant was to be located on a greenfield site?

2  How would you characterize the new human resource strategy in terms of the three ideal types outlined in the introduction to the chapter?

3  Discuss the relationship between the new organizational structure and the wider human resource philosophy and objectives for the new plant.

4  To what extent and why were there inconsistencies and contradictions within the new Aberdare human resource strategy?

5  Assess the strengths and weaknesses of the processes and mechanisms adopted by the company to design the new HR strategy.

6  Discuss the reasons why multinational group management played such a minor role in the evolution of the new strategy.

7  To what extent was the HR strategy as a whole subordinate to wider business and operational objectives for the new plant?

# 7

# The Retreat from Full Flexibility

## Thematic Focus: the Implementation of Human Resource Innovation

In Chapter 5 we reviewed the evidence on the implementation of strategic change and some of the major findings from the literature on the implementation of technical change. In this chapter we will look more generally at the literature on the implementation of organizational innovation and change, with a particular emphasis on the management of human resources.

We will begin with Child's summary (1984: 291–2) of the main factors associated with the successful implementation of organizational change:

1   The change has the support of top management, or at least one influential manager if the change is localized.
2   The change is preceded by careful diagnosis of the problem.
3   There is discussion of the problem, and of ways of dealing with it, with all those affected, and a willingness to adapt plans in the light of this discussion.
4   Different modes of implementation are assessed based on a judgement of the degree of acceptance/resistance among different groups of employees.
5   Training and personal development needs connected with the change are satisfied before rather than after the event.
6   There is a common understanding about the purpose of change, including the role of third parties where appropriate.
7   Mechanisms are put in place to monitor and evaluate the change and its effects systematically.

Most of these factors are self-explanatory. However, two in particular – 1 and 4 – are worth examining in more detail because of their direct relevance to our theme.

Most studies stress the importance of visible and explicit top management support for innovation and change (see McLoughlin and Clark, 1994: ch. 3). They also identify the need for champions of innovation or change agents: to provide continuity between broad strategic aims and their detailed implementation; to drive forward change; and to generate and sustain the cultural framework of change, the shared set of beliefs and values which help harness management and employee support. Where change is being introduced on one site, or particularly in one department or section of an organization, then the champion may well be an individual rather than a team. For Kanter, it is those people who translate strategy into practice, 'and by so doing, shape what strategy turns out to mean' (1983: 54), who are the true corporate entrepreneurs.

Traditionally, textbook advice on the management of innovation and change (see Huczynski and Buchanan, 1991: ch. 20) has stressed the importance of the clear definition of objectives by senior managers and the full involvement of the workforce. More recently, the emphasis has shifted to more manipulative approaches and the often informal ways in which champions of innovation enlist support by explaining and justifying it in ways that will be seen as acceptable. The successful implementation of change in the modern organization appears now to rely less on technical knowledge and expertise than on interpersonal and social skills, on 'communication, presentation, negotiation, influencing and selling' (1991: 544).

However, whichever approach is adopted by champions of innovation, the greatest danger is for them to believe or act as if they are in complete control of events. As McCaskey has written:

> managers who insist they control everything can be devastated when they realise they don't. Also those who lose courage and feel buffeted by the world can't respond effectively. The best coping occurs when managers can bend under the wind and keep stress to a middle level. Tension is high enough to stimulate energy but not so high as to impede performance. This seems to follow from a wise mixture of feeling in control and not in control. (1988: 2)

This reaffirms the point made in Chapter 2 about the need for managers to recognize the multiplicity of goals and objectives in organizations, the innate ambiguity and unpredictability of both internal and external environments, and the need to prioritize certain goals over others at particular times and places.

Another core theme of the implementation literature relates to the different methods of managing innovation and change, in particular the choice between top-down, authoritarian approaches, and participative and consultative ones (see Child, 1984: 281ff.). The general conclusion from past research is that the choice of approach is likely to be contingent both on the seriousness of the problem with which the change is intended to deal (the more serious the problem, the more likely the change will need to be implemented speedily and with only limited time for consultation) and the anticipated degree of resistance to change among different sections of the workforce (extended consultation is likely to be unproductive if some groups are known to be so fundamentally opposed to the change that any attempt at persuasion is likely to be fruitless). All things being equal, however, a participative approach, in which senior managers take employees into their confidence about the reasons for change and consult them about ways to achieve their objectives, offers a better chance of smooth implementation. In the longer term, it also offers a better chance for the organization to develop what Child has called 'an adaptative learning capacity' (1984: 285), the acquired capacity to adapt to change. One of the central themes running through this book is the importance of an organization's 'invisible' (Itami, 1987) or 'intangible' (Pettigrew and Whipp, 1991: 7) assets. These can be nurtured and enhanced during the implementation of change almost irrespective of how successful the change proves to be.

In choosing approaches to implementation, Child has stressed the importance of organizations making a prior assessment of the degree of support or resistance to change among different groups of employees. On greenfield sites, of course, staff will generally be aware, prior to recruitment, of the broad types of technology and employment practices which are to be introduced. Indeed, they tend to be selected or rejected, as we shall see, because of their aptitude or willingness to operate new systems and new ways of working. Nevertheless, the kind of technical and human resource innovations envisaged on most greenfield sites (and certainly those outlined in the preceding three chapters for implementation in the mid- to late 1980s) represent for most employees a radical break from the ways of relating to technology and work to which they have been accustomed.

What evidence exists about likely employee reactions to the introduction of technical and organizational change? The most comprehensive source of data on these issues is to be found in the 1984 and 1990 British Workplace Industrial Relations Surveys of over 2000 companies (for a summary, see Daniel and Millward, 1993). Three general findings from these surveys are of particular interest. First, employee reactions to the introduction of technical change were overwhelmingly positive. Second, organizational change – the introduction of greater work flexibility, changes in staffing levels, pay systems, job definitions – introduced independently of technical change was resisted more often than it was supported. Finally, where organizational change was introduced as a consequence of technical change, it was generally supported. Or, in the words of Daniel and Millward (1993: 56), 'the introduction of new technology smoothed the path of changes in working practice that were less readily accepted in its absence'.

How can this high degree of support for technical change and low level of support for organizational change be explained? Drawing on data from in-depth follow-up case studies from their survey, Daniel and Millward concluded that, for employees,

> new technology represented progress and advance; the benefits of new and improved machines were concrete, manifest and demonstrable; they represented competitive advantage – the modernity of the technology symbolized the standing of a manufacturing workshop to all who knew the industry, both internal and external; investment in new technology represented confidence in the future and hence improved longer-term job prospects and security; many features of new technology were familiar to workers and valued in their domestic and leisure lives; and, finally, the introduction of new technology tended to be an incremental and continuous process. (1993: 62)

In short, technical change symbolized progress, confidence and success. In contrast, organizational change tended to be associated with failure. It represented an admission of poor or inefficient organization in the past. It was seen as a managerial solution arising out of management failures leading to work intensification and job loss for employees without any of the compensating positive features of technical change. On this analysis the combination of a greenfield site and the introduction of human resource changes directly related to the introduction of new technology – as at Pirelli Aberdare – would be expected to provide a powerful structural basis for employee acceptance and smooth implementation of strategic change.

Despite this, resistance to change can arise from many sources, of which Bedeian (1980) has distinguished four main types (for discussion see Huczynski and Buchanan, 1991: 533–4). First, there is parochial self-interest, where individuals or groups see change as threatening vested interests and challenging existing power relationships and arrangements. Second, resistance can come from misunderstanding and lack of trust, whether through the failure of managers to inform employees adequately of the reasons for change or through the provision of incomplete or incorrect information leading to rumour and uncertainty. Third, different individuals or groups may have contradictory assessments of the aims, costs and benefits of change (although, as Child points out, conflicting approaches may also lead to constructive debate and positive alternative proposals for change). Fourth, resistance to change can arise because certain individuals have a low tolerance to change and uncertainty which makes them cling to existing ways of doing things. As we will see, these different types of resistance, in particular the first and the fourth, can be found just as much, if not more, among managers as among non-management staff.

Finally, creating a new organization culture and generating new beliefs and expectations – what Johnson and Scholes (1993: 401) call 'symbolic activity' – is of great importance to the implementation of strategic innovation and change. We have already seen how the pay and grading system adopted for the new Aberdare factory was intended to recognize (and symbolize) the equal contribution of all non-management staff to the success of the enterprise. Recruitment and selection procedures are also a vital mechanism, not only for ensuring the employment of the appropriate staff, but also for sending messages about the nature of the organization and its core values. For example, exclusive reliance on formal interviews with the immediate line manager tends to suggest a hierarchical, bureaucratic and compartmentalized organization, whereas multiple selection tests, more open-ended interviews, and the involvement of more than one manager (including a senior one), suggest a more open and challenging organization (see Johnson and Scholes, 1993: 401–5). The behaviour of senior managers or key change agents can also send powerful messages about the organization, confirming or contradicting its desired image. The importance of symbolic activity, particularly in managing large-scale innovations, will be a major theme in this chapter.

# Recruitment, Induction and Initial Training

The period from May 1987 to December 1988 at Aberdare was dominated, in human resource terms, by the programmes of recruitment, induction and initial training.

## *Initial recruitment*

As we saw in Chapter 6, the four senior managers for the new plant had already been appointed by April 1986. The eight remaining line managers – maintenance manager, chief systems engineer, planning coordinator, accountant, contract sales manager, and the three shift managers – were recruited between September 1986 and June 1987. The major recruitment task for the site from March 1987 was therefore to select the non-management staff, particularly production and maintenance workers. The main administrative functions – sales and accounts – were still being carried out in Southampton.

The first group of non-management employees to be recruited – seven producers and three maintainers – started work in May 1987. All were ex-employees of the old Aberdare factory. During the rundown of the old factory in 1985–6, they had been part of the select few who had been offered jobs in different parts of the company in South Hampshire, travelling down on a Monday morning, staying in digs or hotels, and returning home on Friday night. Some were employed to train staff on the machines which had been transferred from Aberdare to Bishopstoke, others were offered temporary basic grade operator jobs in the Southampton factory, which was running down and where recruitment was proving difficult. Most were initially offered six-month contracts, but for some it turned out to be nearer 18 months. Nevertheless, most had accepted the contracts and the major disruption to their lives in the hope that it would give them a head start in applying for jobs in the new factory.

And so it proved to be. For when administration manager David Lane and operations manager Martin Dawson began recruiting the first group of staff, they soon concluded that the experience, cable knowledge, and loyalty of these employees would be a major asset in the early stages of machine installation and commissioning. During April 1987 they were all interviewed at either Southampton or Bishopstoke and offered jobs without any special recruitment tests. In late May, a second group of four producers and two maintainers were recruited after a presentation and interviews by Lane and Dawson in the local Aberdare job centre.

However, it was the arrival of Mary Johnson as site personnel and training officer in August 1987 which heralded the beginning of the recruitment campaign proper. She was a South Walian by background and had previously been working as a training officer for Asda, the supermarket chain. During September and October 1987 she worked with Lane, Howells, Dawson and the shift managers to devise a recruitment procedure and a set of selection techniques which were to operate more or less unchanged until mid-1989.

Each recruitment round followed a standard procedure. The company put advertisements in the regional and local press and local job centres asking for applications by a certain date for one of three categories: 'producers', 'maintainers' (mechanical, electrical or electronic) or 'administrators' (normally specified more narrowly, such as accounts, sales, or personnel). At the same time, there was a regular stream of unsolicited applications from local people who had heard about vacancies at the new plant. After the closing date, all applications received were sifted through separately by Mary Johnson and one of the line managers responsible (shift managers for producers, maintenance manager for maintainers), and an average of 20 out of the large number of applicants in each round were invited to come to the factory for a whole day. If there were more suitable candidates than places, applications were kept on file until the next round.

At the start of the day, applicants heard two presentations, one from Martin Dawson talking about cable manufacture and the challenge of computer-integrated manufacture, the other from David Lane about the philosophy of the plant and the new personnel, work and industrial relations arrangements. According to staff interviewed in 1990, Lane's presentation made a particularly strong impression. He inspired an enormous sense of optimism, confidence and enthusiasm about the new personnel policies at the factory. At the same time he created high expectations of radically new forms of management–employee relations based on team work, mutual respect and unprecedented levels of two-way communication. As we shall see, some of these expectations were almost inevitably not realized in practice, leading to a sense of disillusion among some of the staff (which would not have been so marked if the initial objectives had not been so ambitious).

Following the presentations all candidates took a series of written and practical tests – including simple computer aptitude and personality tests – to identify their skills and abilities in areas such as new technology, flexible working, team-working and problem-solving. These were followed by a short interview with Mary Johnson and one of the relevant line managers. While technical ability was not unimportant (and a number of producers were technically very well qualified), increasingly the interviewers placed greatest priority on personality and attitude. As one shift manager said: 'If a guy didn't seem to have the right attitude for what we wanted here, the concept of team-working . . . , then in spite of what qualifications and experience he might have, we didn't want him'. The candidate who, when asked about his attitude to self-supervision, replied 'I'll have a slice of that', did not make the next stage! Occasionally, one or two of the applicants withdrew before going to interview, but most stuck the course.

In the days following the interview, Johnson and the relevant line manager(s) discussed each applicant, looking at their test results and their interview performance. Those regarded as appointable were then invited for

**Table 7.1     *Growth of the Aberdare workforce, 1987–90 as at 31 December (no. of resignations in brackets)***

|          | Total | Executive/ senior monthly staff | Producers | Maintainers | Administrators |
|----------|-------|---------------------------------|-----------|-------------|----------------|
| pre-1987 | 5     | 5                               | –         | –           | –              |
| 1987     | 52    | 17                              | 19        | 8           | 3              |
| 1988     | 110   | 24 (3)                          | 39 (2)    | 11 (3)      | 14             |
| 1989     | 162   | 34 (1)                          | 79 (1)    | 25 (2)      | 24 (2)         |
| 1990     | 156   | 28 (8)                          | 87 (1)    | 19 (7)      | 22 (3)         |

a final, longer interview with Dawson and Lane. In most cases between five and eight were taken on after each 'round'. Between October 1987 and November 1988 there were eight rounds in all, taking up nearly all of Mary Johnson's time, and a significant amount of the time of Lane, Dawson and the three shift managers. Table 7.1 shows the build-up of the workforce in 1987 and 1988, which reached 110 by the end of 1988, just five short of the 115 projected by Lane in 1986.

For the managers concerned, the amount of time spent on recruitment and selection was enormous, but it was generally regarded as time well spent. With a few exceptions, the staff chosen were highly cooperative and willing. Indeed, with unemployment in the Cynon valley – from which over 80 per cent of staff were recruited – at around 20 per cent in 1987 this is not wholly surprising. However, Lane and Dawson, neither of whom had knowledge or experience of the Welsh valleys, were struck in 1987–8 by the low morale and low level of self-esteem of many of the applicants. In the previous decades the traditional industries of the valley, particularly coal-mining, had been decimated. At the same time, the number of male manual jobs had also declined substantially, either not to be replaced at all, or to be partially replaced by semi-skilled assembly jobs – traditionally regarded as 'women's work' – such as those at Hitachi's assembly plant 5 km up the road. Pirelli was seen in the valley as the first major company for years to offer 'men's jobs', and many applicants were desperate to get work. For Regan, Lane and Dawson it was both depressing and challenging. It certainly reinforced their view that a heavy programme of training would be vital to prepare staff for the new tasks and approaches to work required in the new factory.

## Induction training

With the support of the corporate personnel department, the Aberdare site was allocated a substantial training budget for 1987 and 1988, and Lane and Dawson – the latter had overall management responsibility for the vast majority of staff on site – worked closely together in the first half of 1987 to devise an overall training plan. They agreed that staff ought to undergo a detailed induction programme prior to entering the plant so that they could be fully prepared for the work that awaited them. After a meeting at the local college early in 1987 they found a kindred spirit in Bob Walker, a 34-year-old Senior Lecturer in Electrical and Electronic Engineering, who agreed to work with them on the induction programme. Walker spent a number of days in Southampton with Dawson and one of the future shift managers learning about cable-making and about the plans for the new factory. He then devised a one-week induction programme, covering everything from the philosophy and business strategy of the plant to cable types, cable markets and hands-on computing. Walker made a trial presentation to Dawson and, when this went smoothly, Aberdare College was awarded the contract to provide induction training – in groups of up to 15 – for all new non-management employees.

These sessions were tailored to the particular groups of new recruits. The first group of ex-Pirelli Aberdare employees, mostly in their 30s and 40s, knew all about cable products and the traditional process of cable-making, but had little knowledge or experience of working with computers. Much of their induction week was spent comparing old and new methods of cable-making and learning hands-on keyboard skills. In contrast, the last main group of producers to be recruited – when the company started to get worried about the rather high age profile of staff, see Technical Appendix for details – were all under the age of 25. They were familiar with computers both from school and home, but had no knowledge of cable-making. Much of their induction course was spent learning about different cable types and the main components of the cable-making process.

Parts of the induction course were well received. Walker was enthusiastic, committed and knowledgeable, if a little loquacious ('he could talk for Wales', as one of the recruits later commented). However, attending a training course for a whole week, from 9 to 4 each day (Dawson had originally suggested 9 to 5), was almost certainly more than most recruits could take, even though formal teaching was broken up with exercises and computer training. Most had not been in a formal learning situation since school, and they were simply unable to concentrate for that length of time. With hindsight there was general agreement from all concerned that it would have been better to have periods at the college interspersed with hands-on experience in the factory rather than an induction course right at the beginning in one block. However, in 1987 and early 1988, this would have been difficult as there was not enough equipment installed and commissioned for all employees to try out their new skills.

### Other initial training courses

A similar problem bedevilled the rest of the programme of internal and external training courses organized until the end of 1988. These included everything from basic skills training – plc training, forklift truck driving pending the commissioning of the AGVs, fire-fighting – to courses on team-building and instructional techniques for trainers. Most of the latter were given by an ex-corporate training manager of PG, who had recently set up his own training consultancy. During 1988, 1132 person days were spent on training at Aberdare, which amounted to around 15 days' training for each member of staff (there were 52 staff at the beginning of the year, 110 at the end). The average number of training days per employee in UK private manufacturing companies at the time was around 5.5 per annum (*Training in Britain*, London: HMSO, 1989: 36).

As we shall see during the course of this chapter, the induction and training programme was highly successful in conveying to staff the kind of new approaches to work – flexibility, team-working, problem-solving – required in the new plant. What it didn't really do was to allow the staff to put the training into practice on the job. When the new PG corporate training manager, Brian Morris, carried out an audit of Aberdare training at the end of 1988, he made two main recommendations to try and meet this problem. First, he suggested that most of the future training should be carried out in house and applied much more directly to specific tasks in the factory. Second, he was highly critical of the failure to make any real progress in specifying the tasks, knowledge, technical and social skills required for staff to achieve skill modules.

# The System of Skill Modules

## *Skill modules: the idea outlined*

The human resource policy which excited perhaps the greatest initial interest of employees at Aberdare – and caused both management and employees some of their greatest problems over the next three years – was the system of skill modules. The Aberdare employment handbook, prepared by Lane and given to every new employee during the induction programme, stated clearly what the company understood by skill modules in the early days:

> [Skill modules] are defined as that combination of skill and knowledge required for satisfactory performance of the activity in question. The work within the unit will be subdivided into a number of different activities or modules, each of which will require combinations of skill and knowledge. There will be a number of 'skill modules' within each of the functions of the unit, each requiring systematic training. At the end of each training you will be objectively assessed to ensure you can satisfactorily apply the content of the skill module before you are awarded the salary increase. (Pirelli General, 1987: 39)

In sum, skill modules were a method of training, a core element of the system of work allocation, and a method of payment. Every member of staff was required to acquire additional skills as determined by management, but taking into account 'both operational requirements and the abilities of the individual'. The handbook went on: 'There will be a limit to the number of additional skill modules that may be attained, and this limit will be determined by management and will reflect the unit's ability to effectively utilize these modules for its operational requirements' (1987: 40).

At his initial recruitment presentations, David Lane expanded on this basic definition. Although he emphasized that there would be a limit to the number of modules that could be attained, he did suggest that employees could expect to achieve between six and ten depending on their abilities and operational requirements. The successful completion of each skill module would attract a pay supplement, set in 1987 at around 4 per cent of basic pay. This was a powerful financial incentive for employees to acquire skill modules as quickly as possible, which was exactly what the company wanted in the initial stages.

Lane also highlighted in his recruitment presentations the company's expectation – supported fully by the corporate personnel department – that all producers, maintainers and administrators would acquire at least one skill module from one of the other two occupational areas. This was felt to be important to the achievement of maximum flexibility in the plant. It was also intended to make employees aware of the wider business of which their own activity was but one part. This idea was met with disbelief by many employees and line managers, but so much of what was being attempted at Aberdare was so novel that they were won over by Lane's commitment and enthusiasm. Illustrations of possible skill module progression for the three main occupational groups were drawn up by the corporate personnel department (see Yeandle and Clark, 1989a, for examples). Producers and maintainers were shown as taking skill modules in accounts, with sales and accounts staff taking modules in packing and winding, and basic mechanics.

In January 1988, the first group of non-management staff was about to complete nine months' service in the new plant, the normal period envisaged for the completion of a skill module. The staff therefore approached Lane via their trade union branch secretary (see Chapter 8) to find out what was happening with the modules. After consultations with Regan and Dawson, Lane prepared a statement which was sent to all staff in February 1988. This reaffirmed the company's commitment to skill modules, but pointed out that they could only be introduced after all the machines and systems had been installed and commissioned successfully.

In the meantime, he announced an interim arrangement whereby every member of staff would be awarded a module after the completion of nine months' service. This would be subject to adequate progress in one particular skill module area, basic competence in the use of VDUs or plc keyboards, and a basic understanding of the site's overall business system (which had been covered in the induction training week). Finally, Lane pointed out that, in the initial period at least, only operators themselves would know in detail about the activities and knowledge required for particular areas of work. He therefore asked them to record this information and pass it on to the training officer (Mary Johnson). The assumption was that she would collate the material, define each module in conjunction with the relevant line manager, and work out a system of recording progress and final assessment.

During the next few months, there was a continual influx of new staff, and no producers or maintainers – let alone administrators, of whom there were only three by February 1988 – had time to record the details of their activities for skill modules. Most were preoccupied in either learning new systems themselves or training new recruits. As we have seen, many of the new machines were still being installed and commissioned at this time. In any event, the whole question became less urgent with the interim solution of awarding an additional module after nine months' service. This is not to say that training was not progressing, particularly in the production area. However, it was now being done on a shift-by-shift basis with individual operators and maintainers passing on the knowledge they had gained, often by example or by word of mouth.

### Lack of continuity in the personnel function

Then in August 1988, totally out of the blue, Mary Johnson resigned to get married and take up a job in the North of England. There was now no specialist personnel manager on site to drive forward the development of the skill module system. Just two months later, David Lane also left the company, having been headhunted by a shipping firm in the City of London (where he had worked prior to joining PG in 1984). This was a bombshell to David Regan, who had formed a formidable partnership with Lane. It was also a bombshell to the workforce, which had had enormous respect and affection for Lane. He had always shown a strong interest in the wellbeing of staff and their families, and had provided a highly effective communications link between management and 'the shopfloor'.

Lane's departure, so soon after the resignation of Johnson, left a gaping hole in the unit's specialist personnel function. Lane had helped to mastermind the new personnel policies at Aberdare and personified their underlying spirit and philosophy, even if he had not always put flesh and detail on them. He was the great communicator in the first 18 months, as well as the general 'eyes and ears' of the factory. Johnson had been instrumental in the whole recruitment exercise and was about to put the skill module system on a firm footing. And now both had gone within two months of each other; one for personal reasons, the other with a lucrative offer it had been difficult to refuse. Both were reluctant to leave Aberdare, but both had compelling reasons to do so.

Graham Howells – who was about to take over as corporate personnel manager on the retirement of Richard Jones – and David Regan had only known for a short while that Lane would be leaving and took immediate steps to find someone to take charge of personnel on site while they sought a permanent replacement. The man they appointed as temporary personnel officer was Michael Jenkins. He proved a highly competent and affable manager. However, his temporary status meant that he could not really make any serious progress on skill modules. Pending a resolution of the question, Regan and Dawson agreed in the autumn of 1988 to make a second skill module payment to all employees after 18 months' service, 'subject to adequate progress' in a second skill module area.

Howells sought initially among existing staff in South Hampshire for a permanent replacement for Lane, but again failed to persuade anyone to move to Aberdare. He therefore decided to advertise the job nationally. In the meantime Regan, Howells and Mancini had reviewed the post and concluded that it should lose formal responsibility for the accounts function and concentrate exclusively on personnel policy and training. The new job was therefore retitled divisional personnel manager. The site accountant, working full-time at Aberdare after his extended period of induction in Southampton, now reported directly to Regan.

In the spring of 1989 Howells and Regan appointed a highly qualified candidate, Andrew Powell, to replace Lane. He had previously worked as a personnel officer with the Welsh-based semi-conductor manufacturer, Inmos, and as a recruitment manager for PA management consultants. Powell's style was quite different from that of the outgoing, idealistic and charismatic Lane. Rather than go out daily on to the shopfloor to speak with staff, he preferred to stay in his office and work on formalizing the details of the new personnel policies.

## Designing and formalizing the skill modules

Skill modules (together with communications, see Chapter 8) were clearly his most pressing priority. After consultations with Regan, Dawson and Brian Morris (the corporate training manager), it was decided to ask Michael Jenkins to stay on for a few months on a consultancy basis with the task of developing the skill modules into an integrated scheme.

It was soon agreed that the definition of the skill modules in each broad area of the factory – production, maintenance, commercial, accounts, personnel – should be undertaken by the line manager responsible. Jenkins would then take their definitions, standardize them into a common format, and prepare a formal document on each module. This would contain: a statement of policy and procedure, an outline of training and testing requirements and skill module activities, a set of self-report log sheets (with tick boxes requiring management approval), a safety declaration, and a final sheet to be filled in by the line manager confirming to the personnel manager that the employee had successfully completed the module. Between April and August 1989, Jenkins and the relevant line managers produced 30 modules. There were to be 8 in production (two for each of the four main machine areas), 8 in maintenance (one for each of the four main machine areas and four 'open learning' modules, of which more below), 4 each in the finance, personnel and commercial areas, and 2 covering production planning.

At the end of August 1989, Powell announced to the workforce that the company was ready to implement the skill modules in all areas. Almost immediately they proved highly controversial. Some of the modules were incredibly detailed, amounting to between 15 and 20 pages of A4. Others only amounted to two or three pages and were extremely general. At both these extremes, the modules reflected the different approaches and commitments of the relevant line managers to the skill module system. Despite the general principle, repeated in the statement of policy at the beginning of each module, that they should all take between six and nine months to complete, there appeared to be a number of inconsistencies, particularly in the production area. Some could clearly be completed in a comparatively short period of time, for example the two in packing and in assembly, whereas others, for example the two in extrusion and in metallurgy, would clearly take considerably longer.

Particularly strong opposition came from the maintenance department. Four of the proposed maintenance modules were linked to 'open learning' packages with learning books (including engineering theory), associated software and written tests. The company had bought the packages after discussions with Aberdare college. Dawson had agreed to provide a room for study on site, allow staff two hours off per week, and pay for Bob Walker from the college to be available on site on Monday afternoons for tutorial support. However, the company expected staff to give up two additional hours of their own time per week to complete the modules (two on electronics, one on microprocessor systems and diagnostics, and one on hydraulics and pneumatics), seeing the open learning packages as general personal skills development as well as site-specific training. While some maintenance staff were happy with this, a small number were only prepared to become involved if all the training time was paid for. They argued that if the company required them to train, then it should pay for it.

Powell decided to sit down individually with all the line managers to see how the various problems might be resolved. In the meantime staff held a union meeting at which they decided to campaign for the module payment to be uncoupled from the successful completion of a test and to be paid automatically on a time-served basis after every six months of completed service. In return, staff would agree to be completely flexible within the factory and to undertake whatever training was required by management. This proposal was rejected by Regan and Dawson in the autumn of 1989 on the grounds that it would mean breaking the link between skill module payments and the proven acquisition of skills. However, they did agree to set up a management–union working party to review the whole question.

Whether a stronger union or employee involvement in implementing the skill module system between April and August 1988 would have prevented these problems will never be known. From the autumn of 1989, however, shift managers began to award additional skill modules to those staff whom they believed possessed the required competency pending the outcome of the working party's deliberations. It was at this time that the system truly began to come into disrepute. No one was coordinating the overall operation of the system. The length of time needed to complete a module varied from four months to one year, depending on the machine area and shift manager concerned. On one shift, in particular, some staff were allowed to accumulate the modules and associated payments extremely speedily. In contrast, others became resentful at not being given the chance to move, either because they were needed to train staff, or because their expertise was needed to resolve major problems in a particular area, or – as some believed – because their face did not fit. Whatever the particular reasons, tying payment to training and work allocation had created a financial incentive for individual employees to move round as fast as possible. This was not always to the benefit of other central aspects of the human resource strategy – thorough training, skilled work, and good team spirit.

## Further changes in the personnel function

Before the skill module working party could meet in the new year, there were two developments in the specialist personnel function. In December 1989, Powell appointed a training and safety officer, Dick Unwin, whose job brought together the training part of Mary Johnson's job and the health and safety part of the job of the purchasing officer, who had left the company. Unwin was a South Walian who had originally trained as a fitter, then moved into training and health and safety, and had ended up working, like Powell before him, for PA management consultants. This was his first job at management level in a company personnel department.

The second development was the announcement by Powell in January that he was leaving Pirelli Aberdare – to join the personnel team at Bosch's nearby factory. This was less of a bombshell to senior management and staff than Lane's resignation; his new job was clearly more challenging than the position at Aberdare. He was to be involved with a new HR director in developing the new human resource policies for Bosch's greenfield site, including negotiating a single-union agreement! However, he had also suffered from conflicting expectations about the job at Aberdare: encouraged by Regan and his senior colleagues to become more independent of corporate personnel, and yet continually having to recognize that many aspects of personnel policy – including the final decision to appoint any new staff and all matters to do with senior monthly staff – were determined by the managing director and the corporate personnel department, with virtually no local discretion.

Once again, there was a gaping hole in the Aberdare personnel department. This was made worse by the fact that the personnel administrator, an industrial relations graduate from Cardiff University who had provided some continuity over the previous three years, had begun six months' pregnancy leave at the beginning of January 1990. The corporate personnel department again sought a replacement personnel manager from among existing staff in Hampshire, but again no one could be persuaded to move to Aberdare. Regan was in favour of appointing Unwin to the job, but corporate personnel was not convinced of his personnel (as opposed to training and safety) abilities and resisted for some weeks. However, eventually they gave in, mainly because PG managing director Martinez was applying pressure on all divisions to reduce staffing levels. However, they did insist that the new position should not be graded as high as it had been for Lane and Powell, partly because of Unwin's lack of experience in personnel, and partly because they believed that the major work on personnel policy, recruitment and training had now been completed.

The new job of divisional personnel and safety officer incorporated within it personnel, training and safety. However, responsibility for the purchasing function was transferred to the site accountant, who was promoted to divisional accountant and became an increasingly central member of the senior management team. Three years after the factory had first opened, finance had therefore replaced personnel in the formal structure of the senior management team, with the strong support of Howells, Barnett and PG accounts manager George Sherfield. Informally, however, Unwin was treated by Regan and everyone else on site as the fifth member of the senior management team.

From February 1990 Unwin had to learn his new job and simultaneously carry out the functions of the site personnel officer. He had little or no time to sort out the vexed question of skill modules. However, by this time, David Thomas had established himself in the new post of manufacturing manager (see Chapter 5) and begun, with Dawson's support, to design a revised system of skill modules for production staff. With Powell's departure, Regan often away on Pirelli business, and the union reps becoming more and more disaffected by the whole question of skill modules, the idea of the joint working party had fallen by the wayside.

### Line managers take over the design of the skill module system

In May 1990 Thomas and Dawson produced the first draft of a new proposal on skill modules for producers. It was rejected unanimously on all three shifts when put to the vote. Two further drafts were produced, and rejected overwhelmingly. Then, in early August, the union branch secretary, who was on sick leave but heard from his shift that negotiations on the latest draft had come to an impasse, declared a failure to agree under procedure and called in Stephen Evans, the full-time union officer from Cardiff. The next day, his three union branch colleagues (who were on different shifts) told him they had actually agreed to continue talking to Dawson and Thomas as they believed there was still some scope for negotiation. The branch secretary immediately announced that he felt betrayed and tendered his resignation.

A further amended proposal was then put to the three shifts and was accepted by a narrow majority of two in mid-August 1990. Straight after this two of the three other union reps resigned. They were fed up with the amount of time they had had to spend on resolving the skill module issue and frustrated at what they felt was the fickleness of some of their colleagues, who on various occasions had pressed for changes which, when accepted by management, were then promptly criticized and rejected.

## Flexibility and Training Outcomes

Before we outline the new skill module system for producers as agreed in August 1990, it is important to understand what had already been achieved by this time in terms of work flexibility and skills training. It will be remembered that all members of non-management staff were contractually required to be 'completely flexible'. This had been specified in more detail in the site employment handbook as an expectation that all staff would undertake 'any tasks or duties within their capability irrespective of function, department or discipline' (Pirelli General, 1987: 14–15).

Much of the recent debate on flexibility has been polarized around two conflicting views of its implications for management control and the experience of work (see Clark, 1993c). What might be called the 'optimistic' view sees increased flexibility as enhancing the skills of the workforce, increasing job satisfaction and generally empowering employees at work (for discussion, see McLoughlin and Clark, 1994: 48–54). The 'critical' or pessimistic view suggests that greater flexibility rarely leads to an enhancement of skills, rather to greater intensification of work and enhanced supervision and management control (on this, see Pollert, 1991b). Against this background, what had actually been achieved at Aberdare by the summer of 1990?

By this time, most producers had acquired between three and four skill modules. This indicated that they had achieved some degree of proficiency on machine processes in at least two of the four main areas of the factory, for example two modules in metallurgy and two in packing, or two in extrusion and one in assembly. When it is remembered many of the machines in metallurgy and extrusion were also tandemized, then it is clear that the range of tasks which a producer could be required to carry out was significantly greater than in more traditional factories, where the principle of one operator per machine predominated. For the maintainers, functional flexibility was even greater, as they were required to carry out tasks in all four areas of production. In addition, while mechanically-trained fitters rarely got involved in anything to do with electronics or computers, they did sometimes carry out electrical maintenance – which would have been unheard of in the old factory – while electrically-trained maintainers also dealt with mechanical faults. Administrators had seen the smallest change in task and job flexibility, tending to carry out traditional tasks – albeit computer-assisted – in traditionally titled clerical jobs such as sales ledger, wages, purchasing and home sales.

As far as job demarcations were concerned, the smallest change was again to be found among administrators. Their jobs were highly circumscribed. They had little task flexibility on a day-to-day basis apart from answering the telephone if a colleague was busy. Two had managed to change jobs by applying for vacancies, but with tight staffing levels and heavy pressure of work, the accounts and commercial managers at Aberdare had been reluctant to encourage the kind of training and job flexibility which would have allowed administrators to gain modules. The majority had only achieved one module in the area in which they were directly employed. Among producers, as we have seen, most staff were capable of operating three or four machine processes, and – as a result of sickness absences, holidays, training, and generally tight staffing levels – normally worked in each of their skill module areas for at least one or two days during any six-month period.

The situation for maintainers was highly complex. None had been required to take on production tasks. However, there had been a substantial relaxation of demarcations between fitters and electricians. At the same time, new demarcations had arisen by 1990, both in work flexibility (as a result of operational requirements) and in pay and grading (arising from external labour market pressures). The original intention had been to employ all-round flexible maintainers on a unified maintainer grade (Band A), and to hire a few electronically trained staff with further education qualifications on the higher grade of auto technician (Band B). However, it had proved increasingly difficult to recruit and retain auto technicians. In direct response to these difficulties, as we have seen, a new post of automation engineer, graded at management level, had been created in the first half of 1989. By this time it had also become clear that basic grade electricians were increasingly in demand, particularly following the announcement in April 1989 that Bosch was about to establish a large new factory 25 km away. When three electricians were lost in quick succession to Bosch and Ford in the autumn of 1989, Dawson and Powell decided to convert all the basic grade maintenance electricians from Band A (maintainer) to Band B (automation technician) following successful completion of a short probationary period.

This created substantial unrest among the mechanically trained fitters, who all remained on Band A. Unrest increased when both fitters and auto technicians refused to accept the revised skill module proposals of the maintenance manager, claiming the open learning package modules also discriminated against the fitters. Quite simply, by 1990 the maintenance department was in disarray, and the problem with the skill module system both reflected and deepened it. The result was that, by mid-1990, maintainers and auto technicians had a smaller average number of skill modules than even the administrators (see Table 7.2).

**Table 7.2** *Average number of skill modules for each occupational group, 31 August 1990*

| | |
|---|---|
| Administrators | 1.1 |
| Maintainers/auto technicians | 0.7 |
| Producers | 3.8 |

*Source*: Personal questionnaires filled in by staff
August/September 1990

## The Revision of the Skill Module System

Given this highly uneven distribution of skill modules, the resolution of the problem had to proceed differently for each occupational group. There was least scope for functional flexibility among administrators. Their rates of pay were relatively high for the locality and unionization was relatively low at just over 50 per cent (see Chapter 8). Some were also already engaged in more general skills training – professional accounting or personnel courses – sponsored by the company. The outcome of all these factors was that nothing was done in 1990 to enable administrators to improve their chances of achieving skill modules. They continued to be regarded as the Cinderellas to whom skill modules did not really apply.

The position in maintenance, where the staff were 100 per cent unionized, was resolved by a 'deal' in November 1990 as a result of which site management agreed to promote all the Band A fitters to Band B, thus re-creating a common salary structure. In return, the maintainers agreed that the skill modules in maintenance should be concentrated in future on the successful completion of the Open Learning packages. They also accepted that these would require out-of-hours study, theoretical knowledge and practical examinations administered by Bob Walker of Aberdare college.

As for producers, Dawson and manufacturing manager David Thomas had come to the conclusion that the system as it had operated over the previous three years needed to be changed in a number of ways. First, they needed to equalize the training periods for producer modules. This they did by increasing the number of modules in the two most complex and innovative areas of the factory – metallurgy and extrusion – from two to three, while retaining two each in packing/rewind and large assembly/small assembly. In this latter area they had originally proposed just one module, but they subsequently increased it to two when they realized that staff opposition could scupper the whole scheme.

Second, as envisaged in the original human resource policy discussions in 1986–7, Thomas and Dawson decided to put a limit on the number of modules which could be taken by any individual in line with what was actually necessary to meet operational requirements. This procedure became known as skill module 'capping'. They decided on a maximum of seven: five to meet operational requirements, plus, pragmatically, the two 'time-served' modules awarded to staff in 1988 and 1989. In addition they proposed that, from then on, each member of staff would be allocated – after individual consultation – one 'primary' and one 'secondary' area of responsibility. For example, some would have primary skills in metallurgy (three modules) and secondary ones in assembly (two modules), while others would have primary skills in packing (two modules) and secondary ones in extrusion (three modules). Producers would be expected to work mainly in their primary area – involving between three and five machine processes – but could be allocated to their secondary area as required. Each employee would be required to work for at least two weeks in every nine months in their secondary area to 'keep their hand in' and to be updated on any machine or process changes. Finally, a common training structure would be established across each shift using redesigned logbooks and minimum training periods for each machine.

## Limited flexibility as full flexibility

Two aspects of this revision are particularly interesting. Management's opportunity to deploy all non-management staff with full flexibility had not only not been used in practice, it was now effectively being limited by the policy of skill module capping. By mid-1990 there had been no flexibility of staff *between* the three main occupational groups, only flexibility *within* them – with maintainers having the most flexible patterns of work, followed by producers, with administrators being the least flexible.

Second, capping at five (plus two 'time-served modules') was not simply a unilaterally imposed management decision, it was accepted as broadly appropriate by nearly 90 per cent (63 out of 71) of producers interviewed in August and September 1990. Those who had opposed the Thomas/Dawson plan in mid-1990 had done so not because they were against capping, but because they had been promised a greater number of skill modules on recruitment (Lane had spoken of between six and ten) and the new scheme did not envisage raising the payment for each module to offset the potential loss in earnings. Also, by this time many staff felt the whole idea had been devalued, resenting what they believed to be the unevenness between and within shifts in the time taken to be awarded a module and the rigour of testing. In fact the shift with the highest average number of skill modules in August 1990 had 4.0, the lowest 3.6, but these averages concealed the wide variations within shifts. Indeed, in one shift there was a clear bimodal distribution of skill modules, with one small group of staff averaging the full seven and the other larger group only two.

From interviews in 1990 with over 95 per cent of all producers and their three line managers, it was possible to identify six main reasons why both managers and staff were overwhelmingly in favour of limiting full flexibility in the production area at Aberdare (the reasons are discussed in more detail in Clark, 1993c):

- The *horses for courses principle*: recognizing that some employees are better suited in their temperament, skills and abilities to certain areas of work rather than others, and that it is in their own and management's interests to allocate them to these areas where possible.
- *Specialist knowledge*: specializing and knowing your job intimately encourages motivation and generates high quality work. Excessive job rotation is demotivating and is likely to produce lower quality work.
- *Ownership of particular work areas*: regular allocation to one main area of work encourages commitment to that area. Between 1987 and 1990 maintainers were not given responsibility for any machine area, and faults tended to be patched up rather than engineered out.
- *Training*: up-to-date and knowledgeable staff who are also good trainers are best retained in their main area of work to pass on their specialist knowledge to others.
- *Skill retention*: if skills are so numerous and complex that they cannot be regularly utilized in practice, then they become difficult to retain and eventually 'unusable'.
- *Tight staffing levels*: if staffing levels are very tight and jobs highly specialized, as in the administration area, it is extremely difficult operationally to spare staff for training in other areas.

Under conditions highly favourable to full flexibility, it had neither been required nor used by 1990. But this does not alter the fact that a quiet revolution had taken place at Aberdare between 1987 and 1990. Compared with arrangements in the old Southampton and Aberdare factories which it replaced, there had been a major task and job enlargement, with most producers covering up to five or six machine processes where they had previously covered only one. At the same time, over 90 per cent of production and maintenance staff interviewed in 1990 showed a complete willingness, within the limitations of their training and capabilities, to be as flexible as management and the operations required. By mid-1990, then, David Regan and his management team had achieved what they set out to achieve, namely full flexibility to do what they wanted. In practice, however, to do what they wanted they did not need full flexibility.

## Self-Supervision and First-Line Management

It has been argued (for examples, see Pollert, 1991a) that increased flexibility in work patterns is often accompanied by increased supervision and management control. On the other hand, there are some prominent examples of advanced manufacturing systems in which the day-to-day management of work is delegated to self-managing work teams supported by a new breed of first-line managers with facilitating rather than controlling roles (see on this Clark, 1993c and 1993e). As one American writer has commented, 'supervision is out, self-management is in' (Dent, 1990: 35).

There was no mention of self-supervision in early management drafts of the human resource strategy for the Aberdare factory or in the single-union agreement, even though it was implicit in many of the company's proposed HR policies. By the time David Lane came to write the employment handbook in the spring of 1987, however, it had become an explicit feature of the new approach to the management of work at Aberdare: 'the unit will function with a "flat" organizational structure which will allow employees to become responsible for their own day-to-day management, thereby creating greater job satisfaction and a stronger commitment to the unit' (Pirelli General, 1987: 21).

In the early days many staff were sceptical of the whole idea of self-supervision. Not only were they doubtful whether their managers would allow it, they were slightly scared of taking day-to-day responsibility for decisions about their work, and convinced that the system would be open to abuse. However, with some encouragement from their shift managers and highly motivated by the trust vested in them, by 1990 the vast majority of staff (over 85 per cent) expressed a strong preference for self-supervision – compared with more direct forms of supervision – as the best way to manage their work. The following are typical of producer comments on the system as of 1990:

> In the old plant it was a case of the foreman was there, the chargehand was there . . . , any problems, go and see the chargehand or foreman and he made the decisions. Although the boys could do it themselves then, you weren't expected to do it, you were expected to go to the foreman or chargehand and ask them their opinion . . . , and they'd go and sort it out for you. . . . That is a real change. The whole unit is based on a self-supervision attitude.

> We do work as a team and it's enjoyable. You come to work on the machines next to one another, and there's no pressure on you . . . I think it's much better working like that. If you've got pressure on you, you tend to rebel against that, you think, 'well why should I?' . . . [The other day] a boy on the other shift came in and he'd left a lot of bobbins out the back of the packing section, just piled up there, and we couldn't move. And I said, 'Well, you move that', and he said, 'It's not my job', and I said, 'What do you mean, it's not your job, we're supposed to be flexible, there's no such thing'. I said 'You belong to Victorian values . . . you're flexible, you can't say it's not my job to move that'.

In these quotations we can identify many facets of self-supervision at Aberdare: the end of demarcations based on the idea of a fixed job; the expectation that individual employees will take decisions previously left to supervisors; the need for self-supervision at team as well as individual level; the absence of an overseeing foreman role; the high level of trust; the problems with traditional attitudes; and, throughout, the contrast with more traditional direct systems of labour management and control. Over 80 per cent of producers interviewed at Aberdare in 1990 said that the responsibility and freedom they exercised in their work – which resulted from the combination of flexible working, multiskilling, and self-supervision – had led to a much greater sense of job satisfaction compared with their previous jobs, even though this often involved intensified work effort over a shift. By

1990 functional flexibility linked to self-supervision was clearly providing joint benefits for management – where it helped the smooth flow of production – and workforce – where it led to increased job satisfaction and interest.

Nevertheless, despite its undoubted success, the system of self-supervision did suffer from abuse. Indeed, during the extended installation and commissioning period from mid-1987 to mid-1989, the combination of non-routinized patterns of working and widespread on-the-job training created an environment in which lax time and work discipline could develop. However, once the factory started to settle into a routine pattern of working, abuse of the system was confined to a very small minority of staff. This abuse took three main forms: poor time management, such as taking long breaks, not arriving for work punctually, and not helping out when an individual's own machine or section was running smoothly; poor work performance, such as reduced work effort resulting in smaller production output, failure to repair or clean machines at the end of a shift, and making wrong decisions when problems arose; and finally unwarranted sickness absence, made worse by the fact that under the single-status agreement this was on full pay. There were even some cases – again well known to the staff, if not to the management – where one or two members of staff would be absent from work without genuine reason during the week, then come in on a Saturday and receive not only their full pay for the week but overtime as well.

Line managers were slow to do something about the abuse, partly because they were unaware of it, partly because they had a strong belief in a 'hands off' approach, and partly because – with hindsight – they underestimated the importance of their disciplinary role in an 'automated factory'. The fact that they failed to tackle the abuse led to a lowering of morale among staff who were working with great commitment to get the factory up and running. It even threatened to undermine the whole basis of self-supervision, because staff felt that if one or two people could get away with it, why should they bother? It was only from late 1990 that line managers, with the support of Dick Unwin, the new personnel and training manager, began taking positive action to deal with the problem, and only in 1992 (see Chapter 8) that it was dealt with more systematically.

Self-supervision at Aberdare also represented a fundamental challenge to some of the traditional functions of the first-line manager (see Clark, 1993e). The shift managers, for example, had a not unusual mix of operational/technical and human resource tasks. The main operational/technical tasks were: setting and monitoring the implementation of operational objectives (weekly work plans, quality standards, scrap levels); dealing with major contingencies (severe machine breakdowns, product defects, competing work priorities); and identifying and acting upon recurrent problems and issues. In the first three years, the first two took priority over the third, whereas, as we shall see, the period from 1991–2 gave much more scope for focusing on the third.

They also had a whole range of human resource management tasks. Some were formal: allocation of work duties on a weekly or fortnightly basis; team briefing (which took place more systematically from 1991); rearranging duties because of staff absence; working out training plans and organizing training; enforcing standards of discipline and self-discipline (this was hardly carried out at all in the 1987–90 period). Other tasks were more informal, such as walking round the factory at the beginning of each shift, keeping an eye and an ear on how production was going, listening to individual problems and suggestions. These informal tasks were not always carried out successfully – and sometimes not at all – in the first years of operation. The absence of a formal appraisal system for non-management staff also meant that there was no real pressure on line managers to monitor the performance of individual employees and discuss problems with them.

This failure to introduce an appraisal system, and the failure to get to grips with the skill module system until late 1990, can be attributed in part to the lack of clear definition of the role of shift manager and the lack of management coordination above shift manager level. But it can also be explained in part by the lack of continuity in specialist personnel management on site. What was missing was someone – or some mechanism – to oversee progress in the human resource area, audit the quality and outcomes of particular policies, and also monitor the performance of line managers in managing their staff. For most staff, their line manager was 'the management' and the main influence on their overall perception of employment relations in the plant.

On a site with a de-layered and lean management structure, the quality of line management – and a clear definition of its tasks – is crucial. This was not fully realized at Aberdare in the initial stages of implementation. Self-supervision does not mean absence of management, but a heightened, more focused and more exposed role for a new breed of line managers.

## Summary

This chapter has reviewed the implementation of the new human resource strategy for the Aberdare plant between 1987 and 1990. It began by focusing on the substantial initial recruitment and training effort, which took up most of the time of specialist personnel managers between 1987 and mid-1988. From then until late 1990, the major human resource issues were the elaboration, implementation and revision of the skill module system, and the modification of the policy of full flexibility of the non-management workforce.

By the end of 1990, the diversity of work arrangements between producers, maintainers and administrators had been formally recognized, and the Aberdare senior management team had retreated from their original requirement for full flexibility. Six reasons were given for the 'retreat': the horses for courses principle, specialist knowledge, ownership of particular work areas, training, skill retention, and tight staffing levels. By mid-1990, the management team at Aberdare had achieved what they had set out to achieve, namely full flexibility to do what they wanted. In practice, to do what they wanted they did not need full flexibility.

After initial scepticism and some abuse of the system, particularly in the early phases, self-supervision linked to functional flexibility clearly proved able to provide joint benefits for management and staff. It also represented a fundamental challenge to some of the traditional functions of the supervisor or first-line manager.

---

## Questions for Discussion

1   Assess the effectiveness of the initial recruitment and training schemes at Pirelli Aberdare, and their significance for overall HR strategy.

2   Evaluate the design and implementation of the skill module system as a method of training and system of work allocation.

3   Analyse the main modifications made to the skill module system in 1990 and evaluate the reasons for their introduction.

4   Discuss the evolution of the role and practice of personnel specialists at Aberdare between 1987 and 1990.

5   What are the connections between self-supervision, functional flexibility and human resource management as operated at Aberdare?

# 8
# Designing and Operating a Single-Union Agreement

## Thematic Focus: 'New-Style' Single-Union Agreements

In Chapter 6 we noted that many of the findings of the research literature on human resource management are both contradictory and contested. However, what is not generally in dispute is that the most significant examples of HRM in practice – in terms of their radical break with traditional personnel policies – are to be found in greenfield site manufacturing companies (see Sisson, 1994b: 23). These have also been influential – despite their relatively small numbers – in encouraging managers on 'brownfield' sites to rethink the future direction of their existing arrangements for managing personnel. While some of these greenfield site companies have been non-union, the majority in the UK have embodied their new human resource strategies in 'new-style' single-union agreements.

It is important to note the term 'new style', because single-union agreements as such are not a new phenomenon in the UK. Indeed, workplaces established in the 1970s (and surviving until 1990) had a higher proportion of such agreements than those established in the 1980s. And of those established in the 1980s, most were not in manufacturing, but in banking, insurance and in retail distribution (Millward, 1994: ch. 3).

'New-style' single-union agreements are those which not only recognize a single union to represent the workforce, but also adopt a range of innovative human resource policies. Five particular policies have been identified as the hallmarks of the new-style 'single-union package' (see Bassett, 1987):

- *Single status*, that is, common basic terms and conditions of employment for both manual and white-collar employees.
- *A company council* for consultation and communication.
- *Functional flexibility* in work tasks and job demarcations.
- A *'no-strike' clause*.
- *Arbitration* as the final stage in collective disputes.

A recent survey of senior managers in 37 establishments operating new-style single-union agreements (IRS, 1993a and 1993b) – 76 per cent of which were on greenfield sites – gives a fuller picture of the new human resource and union recognition policies incorporated in them. Data from this survey are summarized below:

*Human resource management*

92% of respondents said that they expected total or full flexibility from their employees.

76% claimed to have introduced single-status terms and conditions.

40% had a unified (integrated) payment system.

86% said they operated a performance appraisal scheme.

78% used team briefing as a direct form of communication with the workforce, 70% had a company council, and 46% quality circles. All but 4 of the 26 company councils were elected by ballot of both union and non-union employees, and 30% dealt with both pay and general consultative matters together.

86% had disputes procedures providing for third-party arbitration, usually by joint agreement.

46% had explicit no-strike provisions.

*Union recognition and representation*

60% were signed by the two most 'moderate' UK engineering unions, the EETPU (41%) and the AUEW (19%) – now merged to form a single union, the AEEU.

16% were signed by each of the UK's two largest general unions, the TGWU and the GMB.

16% of establishments had experienced attempts by unions other than the recognized union to recruit members.

38% of unions were selected by the company after tendering for recognition in what have become known as 'beauty contests'.

The proportion of relevant employees who were union members varied from 19% to 100%, with an average of 75%.

Over half (57%) of the managers surveyed viewed their relations with their union as 'good', and more than 35% considered them to be 'excellent'.

The respondents surveyed identified four main advantages for management of their new-style agreements: simplified bargaining and consultation processes; improved levels of work flexibility; the avoidance of industrial disputes; and increased employee commitment and involvement. The main potential problem was seen as the threat of recognition claims from other unions. Turning to quantifiable outcomes, only 13% of the companies had experienced any industrial disputes since signing the single-union agreement: annual labour turnover averaged 6.6% and non-attendance rates 4.2%; 60% of respondents regarded their workplace's financial performance as above the industry average; and productivity levels compared favourably with other establishments in the same industry, with 57% of the sample describing them as 'good'. In short, the managers surveyed clearly regarded the HRM practices associated with the new-style agreements as an important factor in securing competitive advantage for their companies. (Since the employee perspective was not reflected in the IRS study, we shall also look below at some evidence as to how employees – at Pirelli Aberdare – view single-union agreements.)

Finally, the IRS survey respondents were asked whether, if they were starting up again, they would still choose a 'new-style' single-union agreement or whether they would prefer a non-union or multi-union plant. All but six (86 per cent) reported that they would still favour a single-union agreement, while the rest said they would choose to operate without unions.

These data present a clear picture from a management perspective of key aspects of human resource policies and practices on single-union greenfield sites. They also demonstrate, as we shall see, that the policies contained in the Pirelli Aberdare agreement were broadly typical of the generality of such 'new-style' agreements in the 1980s.

Companies which choose to establish new plants on greenfield sites have a unique opportunity to consider strategically what kind of relationships they want to have with their workforce. Among the choices they face are whether to adopt some variant of human resource management and also whether to recognize trade unions (see Wickens, 1987; Trevor, 1988). On this latter point the evidence suggests that questions of ownership and nationality of the company are likely to play a crucial role. Guest (1989: 48) has argued plausibly that American companies, coming from an HRM tradition of anti-unionism and individualism, are likely to pursue a non-union path. In contrast, Japanese companies with their tradition of 'house' unions are likely to seek some kind of non-union company council or a comprehensive single-union agreement. Finally, European companies more accustomed to pluralist systems of labour management (such as Pirelli) are more likely to opt for single-union recognition and some form of collective bargaining alongside new style HRM policies.

---

# The Making of the Agreement

## *Consulting local industrial relations professionals*

In early 1985, as we saw in Chapter 3, a member of the Aberdare project team had met Peter Edwards, General Secretary of the Wales TUC, to discuss the question of union recognition. This meeting was so secret that even corporate personnel manager Richard Jones did not know about it. It was not until August 1985 that Jones came to discuss these questions with Edwards, in a meeting at ACAS offices in Cardiff arranged by a member of the ACAS Work Research Unit. The meeting was also attended by the Director of ACAS Wales, Keith Abell, and Mark Ball, divisional personnel officer at the existing Pirelli Aberdare site.

Jones began the meeting by outlining the company's plans for the new factory, emphasizing the need for full workforce flexibility and a new approach to industrial relations. He then asked Edwards and Abell for their views on union representation. Both drew attention to the high level of unionization in South Wales manufacturing companies, and Edwards argued that if Pirelli wanted to go down the non-union route, it would almost certainly be faced with pressure for union recognition. Abell pointed out that this would be a high-risk strategy in South Wales, where most inward investors had opted for a comprehensive single-union agreement.

Edwards confirmed this, saying that the Wales TUC had two overriding objectives in dealing with inward investors: to secure the investment (and therefore jobs) for Wales, and to achieve union recognition. If Pirelli was prepared to meet these two basic objectives, the Wales TUC and its affiliated unions would be able to meet most if not all of the company's work and industrial relations requirements. He also emphasized that it was extremely important that the procedure for selection of the union should be open. It should certainly not follow the 'Heathrow agreement' approach, so called after a celebrated case in which one particular union had received advanced notice of a Japanese company's intention to locate in the UK and had secured recognition rights at London's Heathrow airport before any other union even knew about it. If Pirelli played it fair, then Edwards guaranteed that no unions other than those chosen by the company would attempt to recruit or press for recognition.

They then discussed the sensitive question of 'bridging the gap' between the old and the new factories. Jones came away from this part of the discussion with an unmistakable conclusion. If PG decided not to close its old Aberdare factory (this decision had yet to be taken, see Chapter 3), but to keep on the payroll a few of the most trusted employees to provide continuity between the old and the new, it would almost certainly make it much more difficult to achieve a single-union agreement. Under such circumstances, the four unions who still had members at the plant – at that time, the GMB with 40 production workers, TASS with 20 clerical and technical staff, AUEW with 6 mechanical craftsmen, and the EETPU with 1 electrician – would have a strong moral claim for recognition, if not for bargaining purposes, then at least for rights of representation in cases of individual disputes. The result would be the continuation of multi-unionism. Making a clean break, and setting up a new factory on a greenfield site, would clearly provide fewer constraints on the development of new union recognition and representation arrangements.

### Opting for single-union recognition

The rest of the meeting was spent exploring various options and procedures for union recognition. Edwards argued that Pirelli should consider other options apart from a single-union agreement. He suggested single-table bargaining (see Marginson and Sisson, 1990) or even the recognition of one main union for bargaining purposes, together with voluntary membership of other unions with rights of representation in individual cases. However, most of the time was spent examining the single-union option, in particular which unions should be approached and how to proceed in selecting them. Edwards suggested two main criteria: those which were most appropriate to the type of work and skills required, and those which were already represented at the existing/old site. While he could not be involved in any direct overtures to the unions, he would be available to comment on any material prepared in connection with the selection of the union. Edwards concluded by arguing that achieving a fully flexible workforce would require substantial company investment in training. It would also require harmonization of basic terms and conditions of employment for manual and white-collar staff (single status).

The meeting confirmed Jones in his view that the conclusion of a new-style single-union agreement was the best option. However, he had a problem. Unit manager-designate David Regan was not in favour of union recognition, believing that it would provide an alternative focus of loyalty and also inhibit flexibility on the shopfloor. Over the following months, Jones continued to impress on Regan the pragmatic reasons for going down the single-union route. If they decided for the non-union option, they could lay themselves open to months or even years of union pressure and the eventual recognition of unions they might prefer not to recognize. They now had the opportunity to select the union of their choice together with a guarantee from the Wales TUC that if it was done fairly and openly, no other union would have a legitimate reason to complain. These arguments were very similar to those which led Nissan's personnel director Peter Wickens to opt for a single-union agreement at the company's Sunderland plant just a year or so before (quoted in Bassett, 1987: 151): 'We considered the alternatives of no trade unions, and a multiplicity of unions . . . We rejected the first because it would lead to several years of counter-productive antagonism, and the latter because sooner or later it would lead to an erosion of our flexibility and single-status objectives.' Somewhat against his better judgement, in the spring of 1986 Regan agreed to accept the single-union option.

As a result of his friendly disagreement with Jones over the question of union recognition, Regan had become acutely aware of the importance of the corporate personnel department in the multidivisional Pirelli company. He realized that it was simply not possible to adopt new human resource policies for Aberdare which did not have the active support of corporate personnel specialists. On 10 April 1986, therefore, just 10 days after Aberdare administration manager David Lane took up his new job, Regan wrote to Jones suggesting that the time had come to be 'more specific' about the personnel policies for the new site. In order to do this, it was necessary 'to ensure that we are clear on what Pirelli General policy currently is or can be'. It was this initiative which led Jones, Regan, Lane and corporate employee relations manager Graham Howells to form an *ad hoc* working group on personnel policy for the new site, and Lane to write an initial draft of a paper on the new HR policies (discussed in Chapter 6). The first draft of 30 April 1986 stated unequivocally that the company intended to go for single-union recognition.

### Choosing the five 'appropriate' unions

Lane's paper of 30 April outlined in some detail how the company might go about 'appointing' the union. Using the jargon of commercial contracting, Lane proposed that specific unions should be invited to 'tender' for recognition in accordance with a 'specification' which would include the overall philosophy of the unit and a statement about the need to avoid industrial action. At this stage, he explicitly rejected the idea of a no-strike clause, arguing that this would require some kind of binding arbitration. He doubted whether the management of one particular unit within the multi-site Pirelli company would be able, or wish, to give such an undertaking.

On 1 June 1986, Howells took up his post as employee relations manager. He immediately began to work with Lane to refine the draft tender letter to unions together with an accompanying statement of objectives, philosophy and principles. Having to prepare such a statement forced the company to be systematic, explicit and open about their personnel philosophy and policies. No such systematic and comprehensive statement had ever been drawn up for another Pirelli General plant.

On 8 September 1986, after four separate meetings at which comments were received from Jones, Regan, Abell, Edwards and Stanley Russell of the ACAS Work Research Unit, a final version was sent to five unions: the AUEW, EETPU, GMB and TASS (all had previously been recognized at Aberdare) and the TGWU, the company's main union in South Hampshire and a strong presence in manufacturing in South Wales. A preliminary response was requested by 26 September. The letters were signed by Howells, who was now leading the company's negotiations.

### The beauty contest

All five unions responded positively by 26 September. The GMB sent a 10-page draft recognition and procedure agreement for consideration. The EETPU and AUEW sent copies of their single-union agreements with other South Wales companies, while TASS was also positive and encouraging. The TGWU, in contrast, spent most of its reply reasserting its belief that it had to a right to recognition because the new jobs at Aberdare were being transferred from Southampton, where it was the main recognized union.

At the beginning of October, Howells arranged meetings with each of the unions (the 'beauty contest'), to be held at the Crest Hotel, Cardiff on 29 and 30 October and 4 November. He also telephoned Edwards to update him on events. In the meantime, he and Lane prepared a first draft 'Heads of Agreement' for the new unit, which they tabled on the day as a basis for discussion with each of the unions. This contained details of representation arrangements and payment of union subscriptions, and also, for the first time publicly, the company's view on pay negotiations and disputes procedures. Pay would be negotiated by a joint committee of internal union representatives and local plant managers. Only if this failed would the Aberdare divisional manager and a full-time union officer be brought in. If they also failed to agree, the matter would be referred to binding arbitration under the auspices of ACAS. This was intended to 'preclude the need for recourse to any form of industrial action'.

This was the first mention of a 'no-strike' element in the agreement, and was contrary to the sentiments expressed in Lane's initial paper of April 1986. In fact, Jones and Howells had never been against the idea of binding arbitration in principle, but they were suspicious of pendulum arbitration, which was common in agreements made with the EETPU and appeared to reinforce a 'win–lose' approach to disputes rather than the more 'positive sum' approach represented by conventional arbitration. In fact, the first pendulum arbitration case in the UK – at Bowman Webber, the mirror and glass manufacturer – had just been decided (in the summer of 1986) and its complexity only served to confirm their worries (see Lewis, 1991). On the other hand, while a no-strike agreement would certainly not guarantee that there would be no strikes – an impossibility in a free society – it would be symbolic of the new approach to industrial relations and human resource management desired by the company. Jones and Howells were thus able to persuade Lane that they should write in a no-strike clause with conventional binding arbitration as the final stage.

Each union was given a half day to present their case to Jones, Howells, Regan and Lane. The AUEW, EETPU, TASS and GMB were represented by local and national officers, while the TGWU sent a regional organizer and a research assistant. All the presentations were done in a highly professional manner, and over the next three months, the personnel subgroup had lengthy internal discussions about which union to select. By early February they had reached a preliminary decision. They would reject three and continue more detailed discussions with the other two.

### The reasons for rejecting four of the five unions

A confidential personnel department briefing paper – written in February 1987 at the request of sector general manager Arthur Stokes (see Chapter 6) – gives a unique insight into the reasons for this decision. The TGWU was rejected for three main reasons: its generally 'unenlightened attitude' and lack of experience in dealing with new-style agreements; its unwillingness to accept the need for compulsory arbitration and a no-strike clause; and the company's desire to 'isolate' the Aberdare factory from the industrial relations activities of its principal trade union in Hampshire. As for TASS, the briefing paper argued that it had no experience at all of such agreements; it was also opposed to compulsory arbitration. In addition, mention was made of the 'well-documented left-wing attitude' of the national union and its officers (its then General Secretary was a prominent member of the Communist Party of Great Britain).

The AUEW was assessed more positively, but ultimately rejected because its organizational structure was felt to be unsuitable: it was an almost exclusively blue-collar union in an environment where it would also need to represent white-collar staff. Its emphasis on district committees – rather than workplace branches – as the main forum for union decision-making might also allow industrial relations to be affected by external issues and influences. This explanation, together with the company's insistence that negotiations should take place without any reference to outside (national) agreements, make it clear that it was aiming for a self-contained plant-based system of industrial relations for non-management staff at Aberdare.

As for the EETPU, it was clearly the most successful union in Wales – and in the UK more generally – in concluding new-style single-union agreements. However, the briefing paper noted that 'this made them unpopular with other trade unions and thereby creates controversy'. Over the previous six months Jones, Howells and Lane had been told informally on more than one occasion that if the company decided to recognize the EETPU, this would be – as one informant described it – 'like a red rag to a bull' for unions in South Wales, as well as for the TGWU southern regional office which covered the South Hampshire factories. They also learnt through the South Wales grapevine that the EETPU was not expecting to win recognition. On top of this, the internal briefing paper noted that the EETPU often pressed for a 'higher profile with organizations than managements might sometimes prefer'. In other words, the EETPU would be a substantial presence in the factory and extremely vigilant in ensuring that agreements were kept and procedures followed.

### Selecting the GMB

According to the personnel department briefing paper of February 1987, the GMB – and its managerial, administrative and technical staff section, MATSA, which had by now taken over responsibility for the negotiations with PG – was recognized as an organization which, since the election of general secretary John Edmonds in 1984, was 'consciously and successfully' making a name for itself as a progressive and forward-looking union. It had also shown itself to be the keenest of all the unions to be selected and therefore 'very accommodating' in meeting the company's requirements. Its Welsh region also had a policy and attitude of 'non-interference and moderation'. In other words, the company would be likely to see little of the GMB full-time union officers and generally have an easier time of it than with the EETPU. The briefing paper concluded that selection should be determined ultimately by the 'realistic relationship' that could be achieved between the company, the union and its officers 'as they manifest themselves and operate in South Wales'. Thus, while national policies and personalities were undoubtedly important, the character of the South Wales unions and industrial relations scene played a decisive role in the eventual decision.

On 19 and 20 March 1987 Jones, Howells and Lane had further meetings with the EETPU and MATSA. However, the comments in the February briefing paper show that the die was already cast for MATSA. A final meeting was held on 16 April between Pirelli managers (Jones, Howells, Lane, Regan and Dawson) and MATSA officials from South Wales, at which a few minor modifications were made to the final text of the agreement. In essence, though, this was not really a negotiated agreement, but the acceptance by the union of a *fait accompli* in return for recognition. A press release was issued by the company on 24 April announcing the conclusion of the agreement. It was signed at a formal ceremony at the Aberdare plant on 8 May 1987.

### Setting initial pay levels

By this time, 10 non-management employees had already started work in the new factory. In anticipation of this, Lane had been gathering data on pay rates in South Wales throughout March and early April 1987. He commissioned a survey from a prominent firm of industrial relations consultants, acquired information from the local Aberdare job centre, and also carried out a telephone survey of 11 local engineering companies (including Sony, Hitachi and AB Electronics in the Cynon valley, and Hoover in the neighbouring Merthyr valley). He had also updated the 1986 pay rates in the old Aberdare factory, comparing them with the rates paid for non-management staff in Pirelli's Hampshire plants.

Regan was keen that Pirelli should be a market leader in pay, in line with the progressive technical and human resource image he was trying to project for the site. However, he was persuaded by Jones and Howells that the initial basic salary should not be set in the top decile of manual earnings in South Wales, as it would be possible for most employees to reach this level within three to four years once they had acquired 'skill modules' for various work areas. Each skill module was set to attract a payment of £250, around 4 per cent of initial base salary. The initial pay rate set for the basic grade (Grade A) was 17 per cent above the average for the area for production staff, 20 per cent above the average for the mainly female clerical and administrative staff, but 10 per cent below the average for craft-trained maintainers. The personnel working group decided to get round this latter problem by awarding all maintainers who had completed a craft apprenticeship immediate payment of a 'skill supplement' – amounting to five skill modules – on top of their basic pay.

In effect, therefore, the Aberdare site started with four 'basic' grades: Grade A for all producers and most clerical/administrative staff; Grade A plus skill supplement for the majority of maintainers; Grade B for a few higher grade administrative staff with vocational training qualifications; and Grade B plus skill supplement for time-served electronic technicians with vocational training qualifications. The manipulation of the skill module and grading system to meet particular internal and external labour market problems would become a regular feature of industrial relations over the next few years. The pay levels set unilaterally by the company in April 1987 were fixed for 18 months.

### The main features of the single-union agreement

The new-style agreement signed in May 1987 between Pirelli General and the GMB had the following main features:

- *Philosophy*: an explicit statement on the objectives and operational requirements of the plant and its management style (see Chapter 6 above).

- *Coverage*: all non-management employees, but excluding managers and professional specialists.
- *Work requirements*: 'full flexibility' across the three main functions of production, maintenance and administration.
- *Single status*: most notably, the right of all employees to four months off work on full pay, and two further months on half pay, for certified sickness – a major advance on the previous arrangement at Aberdare where manual workers were only entitled to statutory sick pay plus a modest level of company sick pay.
- *Integrated salary structure*: only two basic grades, plus an allowance for shift workers together with skill-based increments for all.
- *Training*: an explicit management commitment to continuous training for all staff via a system of skill modules, with payment for the acquisition of new skills.
- *Recognition*: a single union recognized for both representation and bargaining purposes.
- *Pay determination*: to be conducted at establishment level by representatives of management and union without the presence of the full-time officer, except in cases of failure to agree.
- *Collective disputes*: 'no-strike' clause with binding arbitration as a final stage in cases of failure to agree.
- *Consultation*: direct consultation with the workforce through regular team briefings, monthly 'business review' meetings, and three-monthly general meetings.
- *Grievance and disciplinary procedures*: mixture of union-based (grievances) and non-union-based (discipline) representation; arbitration (grievances) or unilateral management decision (discipline) to be the final stage in cases of failure to agree.

These features of the Aberdare single-union agreement were similar to those outlined in the first part of this chapter. However, the three-page section at the beginning on the general philosophy and operational requirements of the plant (see Chapter 6), and the interrelated provisions on salary structure, skill-based pay and training requirements, were particularly novel. In addition, local union representatives were given a strong and exclusive role in pay bargaining, unlike, for example, at Nissan's Sunderland plant where pay and consultation issues were dealt with together by elected representatives of the whole workforce, irrespective of union membership.

## A rather awkward postscript

The conclusion of the agreement with MATSA in May 1987 had a rather awkward postscript. At the Wales TUC in late April a bitter public row had broken out between the EETPU and the TGWU over a single-union agreement at the Yuasa Battery plant at Ebbw Vale in one of Aberdare's neighbouring valleys. Underlying the dispute, in which both unions accused each other of poaching, were a number of more fundamental issues. The EETPU was on the ideological right of the trade union movement and, given the anti-union climate of the time, was prepared to be pragmatic and give up the right to strike in particular companies in order to win recognition. In contrast, the TGWU, ideologically on the centre-left in the TUC (although on the centre-right in Wales), argued strongly against 'no-strike' agreements, arguing that they amounted to a denial of an historical right of free and independent trade unions. Whatever the differences in principle, the upshot was that the EETPU was highly successful in winning 'beauty contests' for recognition, and the TGWU generally unsuccessful.

During late 1986 and early 1987, GMB general secretary John Edmonds gradually came to ally himself with the position of the TGWU. In one particular speech he argued:

> The most objectionable part of no-strike deals is that very often they are signed when nobody is employed. They are signed on a greenfield site when the union bureaucrat is representing three blades of grass and a lump of concrete. People are then recruited into that organisation against a background of four million unemployed and told that they can't take strike action whatever the circumstances. (Cited in Bassett, 1987: 177)

This is almost exactly the kind of agreement made by the Cardiff regional office of the GMB at Pirelli Aberdare. Sensing a good story, a number of local and national newspapers contacted Pirelli and the GMB following the signing of the deal on 8 May 1987 to find out whether the agreement included a no-strike clause and/or binding arbitration. Neither the company nor the South Wales region of the GMB wanted any adverse publicity for the deal, so they agreed to keep its terms confidential. However, on 19 June the labour correspondent of the *Financial Times*, Charles Leadbeater, contacted MATSA's national industrial officer, who had been involved in the negotiations. He confirmed: 'It does contain a no-strike deal'. On 20 June this was duly reported in the *FT* by Leadbeater, who pointed out that this would be likely to undermine John Edmonds's efforts to get the TUC to outlaw beauty contests and strike-free agreements.

A few weeks later, Graham Howells received a telephone call from a rather embarrassed Cardiff office of the GMB to see whether he would be prepared to look again at certain clauses in the agreement. On 4 September 1987, he travelled down to Cardiff to meet the four GMB officials who had been involved in the deal. They suggested a rewording of the paragraph on the preclusion of industrial action to avoid explicit inclusion of the no-strike clause. In return they were willing to confirm to Howells privately in writing that that was indeed their interpretation of the meaning of the paragraph. Howells responded that this paragraph was by no means the most important part of the agreement in the company's eyes. However, it was symbolic of the new approach to industrial relations at the new site, and it had been jointly agreed only five months previously.

No agreement was reached, and both parties agreed to consider the matter further. The GMB regional secretary Eddie Sinfield, only one year from retirement and an imposing and authoritative figure, was clearly not happy at what he saw as the interference from head office in a hard-won recognition deal. He decided to take soundings among the workforce at the Aberdare site. There he found almost unanimous support for the preclusion of strike action at the factory (most of the initial recruits had previously worked at the old factory and did not want a return to old-style adversarial industrial relations). Sinfield wrote to the GMB head office passing on this information about the feelings of the membership and asking for instructions as to what he should do. He never received a reply.

## Implementing the Agreement

### The formation of the union branch and the 1988 pay negotiations

The workplace union branch at the Aberdare plant was established in November 1987 following a visit from two full-time union officers based in Cardiff. At this meeting union members also elected a branch committee of four, a branch secretary (an ex-miner) and three other members. In the initial months there was very little for the committee to do apart from recruit members. Lane and Dawson agreed to give them a slot in the induction programmes for new staff, and by the middle of 1988 they had achieved 98 per cent membership of non-management staff. After pressure on Lane they also secured a time-based payment for the first skill module in early 1988 (see Chapter 7).

The first major union initiative came in August 1988 when the branch secretary called a general meeting to discuss the first pay claim – the pay rates were due to be reviewed by 15 November. In time honoured fashion, the branch committee prepared a 'shopping list' of nine demands – including a review of the skill module system – and these were agreed and presented to the company in September 1988. After a preliminary meeting with Lane to discuss procedure, Lane and Dawson met the local committee in early October and offered an increase in basic pay of just over 7 per cent, around 10 per cent on each skill module, and 11 per cent on the shift premium (which of course did not apply to administrators). The total cost to the company came out at just above the current annual increase in the retail price index. The offer was rejected by the staff at a general meeting held on the same day.

The next negotiating meeting, one month later, was chaired by David Regan (in the interim David Lane had announced his resignation from the company, see Chapter 7). Regan began by expressing his grave disappointment at the old style of bargaining, with the union putting forward a shopping list, the company responding, the union rejecting, and so on. He expressed the hope that this could change next time. Nevertheless, he did present a revised offer, raising basic pay to around 9 per cent, keeping the skill module increase at around 10 per cent, but raising the shift allowance by around 30 per cent (from £1300 to £1700). This was conditional on the staff accepting that the deal would last for 15 months. This was again rejected at a union meeting. Finally, a few days later Regan proposed, and the staff accepted, that the deal should run for 13 months, from 15 November 1988 to 15 December 1989.

## New communications systems

'Business review and consultative meetings' had originally been seen as the centrepiece of the factory communications system, along with team briefing and quarterly general communications meetings. In conception at least, the business review meeting (as it came to be called) was the most radical and innovative of the three. Its constitution as laid out in the single-union agreement envisaged regular consultation and information provision about the trading position of the factory, its efficiency and performance in meeting targets and objectives, and matters of health and safety. In fact, not one business review or communications meeting was held in the first two years after the opening of the factory. Team briefings did take place in some areas, but they tended to be irregular meetings between line managers and their staff called at short notice to discuss particular operational or organizational problems.

In the 'extended honeymoon' period, the number of staff was so small, and the interaction between staff and their managers so regular and direct, that formal communications had not seemed necessary. However, by April 1989, there were 130 staff in post, and Lane's replacement as personnel manager, Andrew Powell, persuaded Regan that it was an appropriate time to institute the business review meetings. The first meeting consisted of the senior management team, the four union branch representatives, and one elected member of senior monthly staff (there should have been two according to the constitution). Regan suggested to Powell that the meetings should take place immediately following the monthly executive managers meeting (EMM) at which he and the three other company divisional managers reported to the PG managing director on the performance of their division in the previous month. Regan also felt that the agenda of the Aberdare meeting should follow very closely that of the EMM, beginning with a review of developments in the company as a whole, then concentrating on the performance of the Aberdare factory – its sales, output, profitability and personnel matters.

The meetings were a revelation to the union representatives, and to most of the senior managers too. Regan seemed to hold back little in terms of information. He would talk openly about everything from the policies of PG's main competitors to the impact of interest rate changes on profitability, often spicing his contributions with anecdotes which gave those present a real sense of being 'on the inside'. The authenticity of his performance was underlined by the fact that he used the same overheads that he had used for the executive management meetings. Regan was practising what he preached about openness, and soon the business review meetings were opened up to additional members of staff who wished to attend. Attendance between 1988 and 1990 averaged around 15, roughly half of whom were non-management staff.

At the second meeting in June 1989, Regan announced that the company had decided to spend over £1 million on a new 'compound' plant on the Aberdare site to manufacture PVC compound for cable insulation. It would replace the old PVC factory on the Southampton site, and create between six and nine new jobs at Aberdare. Further review meetings were held in September, October and November 1989 and January 1990. In each case, Powell took responsibility for organizing the meetings. He also prepared an agenda and wrote up summary minutes which were posted on the noticeboard. This degree of formality was quite an innovation for the site, where Regan had established a highly anti-bureaucratic culture, with no names on doors, no minutes of meetings, and no personal secretaries for senior managers. By the beginning of 1990, the business review meetings had become an established and highly innovative part of the internal communications system at Aberdare, albeit attended by only a small minority of the workforce. It had, though, needed the arrival of a new specialist personnel manager to ensure that the meetings got off the ground.

## A New Style of Pay Negotiations

As we have seen, Regan had not been at all happy with the 1988 pay negotiations. In the autumn of 1989 he proposed to site union representatives that they should try and conduct the 1989 pay review on a more high trust basis. Now that the business review meetings were up and running and he had given them detailed information about the unit's performance and its various targets, Regan proposed that each pay review should begin by a joint management–union review of the year's performance together with a discussion of the coming year's management plan. (The 'Manplan' was the site plan for the coming year, agreed with the managing director, containing targets on everything from output and scrap levels to sales per employee, as well as assumptions about total earnings increases.) Regan and Powell would explain how the year's pay assumptions had been arrived at, and then discuss jointly with the union reps how the overall figure would be distributed between basic pay, shift allowance and skill modules.

The union reps were unsure about this proposal, even more so as they had just lost their experienced branch secretary who had resigned from the company. However, they had been impressed with Regan's openness over the previous few months, and were prepared to give it a try, on condition that they should be given training in company accounts and how to read balance sheets. This was agreed (but never happened).

Negotiations were conducted on three afternoons over the period of a week in late November 1989. After Regan had gone through the Manplan targets, Powell announced the overall figure agreed between Regan, PG managing director Martinez, and company personnel manager Howells. It was to be 10.1 per cent. This was greeted with visible signs of relief from the union reps. It was around one percentage point above the retail price index and they knew it would be acceptable to staff. In fact, they had not been expecting to get much of an increase at all, as the factory was still a long way from making a profit. Powell then explained how the salary assumptions had been arrived at, stressing that they had been able to achieve a relatively good deal – slightly higher than in the rest of the PG factories in South Hampshire – because they had persuaded the MD and the corporate personnel manager that they faced strong competition for skilled staff from Bosch and Ford (see Chapter 7).

Powell proposed that the 10.1 per cent should be distributed in roughly equal percentages across all three elements of the pay structure. Given that all maintainers were already paid an automatic skill supplement for their craft training and all were working shifts, this proposal clearly favoured maintainers most – the group for whom there was greatest competition in the labour market – and administrators least – who were already being paid relatively well in comparison to local rates. After consultation with staff, the union reps proposed that, as the previous year's increase had heavily favoured those working on shifts, the whole of the 10.1 per cent should go on basic pay, with nothing at all on the skill modules. Indeed, to impress on the management the staff's negative attitude towards the way skill modules had been handled (see Chapter 7), they proposed the introduction of flexible summer holidays in 1990 rather than having a two-week shutdown, even though Regan pointed out that this would create almost permanent staff shortages between June and September and would thus completely interrupt the skill module training programme.

After further discussions, in which Regan and Powell made clear that they had only been able to secure the 10.1 per cent increase from Martinez on the grounds of local labour market problems, the union reps agreed to distribute it equally across all three elements of the pay structure. Clearly Regan's original conviction – that all non-management grades should be paid the same in recognition of their equal contribution to the work of the factory – was gradually being undermined.

The union reps did achieve two successes, however. They were so angry at the delays over implementing skill modules that they persuaded Regan and Powell to agree to flexible summer holidays (between June and September) rather than the normal two-week shutdown in August. They also secured an agreement to defer payment of the small nominal increase per skill module (£25) until a system was agreed that was acceptable and fair to all. This was eventually achieved in late 1990, but not before disagreements within the union negotiating team had led to the resignation of three of the four branch officers in the summer of 1990 (see Chapter 7).

In the autumn of 1990 an almost brand new branch executive, now led by a respected and consensually-minded producer who had worked in the old Aberdare factory, successfully negotiated the 1990 pay round, achieving a settlement broadly in line with the annual increase in the RPI. Over the following year the union branch had a comparatively low profile, dealing with normal run-of-the-mill union issues – health and safety checks, voicing occasional grievances, chasing up claims for those injured at work, and also representing the first two staff who were given disciplinary hearings because of timekeeping and absence irregularities. However, this period of calm proved to be the lull before the storm.

# Changes in the Industrial Relations Climate

The first signs of impending change came with the reorganization of PG in July 1991 into two main divisions (for details see Chapter 9). The Aberdare building wires factory was brought together with two other sites – industrial cables and power cables – to form a new Energy Cables Division, with its own divisional managing director (Keith Parsons) and personnel manager (Elaine Jenkins) based at Eastleigh. It also had a new divisional commercial director, also based at Eastleigh, with responsibility for sales and marketing and covering all three of the previously separate divisions. This new post was offered to David Regan, who left Aberdare in September 1991 to take up his new position. His old job – now called site manager rather than divisional manager – was filled by Michael Elliott, who had been working at Aberdare since September 1990 as operations manager (see Chapters 9 and 10 for further details of these changes).

## Changes in communication arrangements

The creation of the new Energy Cables Division (ECD) acted as a catalyst for a change in communication and consultation arrangements. Since mid-1989, as we have seen, the managing director Felipe Martinez had chaired a regular monthly review meeting at Aberdare to monitor the progress of site operations. From now on, monthly monitoring of the performance of individual sites was carried out at ECD level, with all site managers in the new division reporting to a meeting chaired by the divisional managing director Keith Parsons. The existing operations review meetings at Aberdare continued to take place, as they had proved useful in bringing together all the line managers on site to discuss performance, common problems and future plans. However, in the absence of senior corporate managers from outside Aberdare, the site review meetings now overlapped substantially both in content and membership with the business review meetings (with the exception of the union and employee representatives).

Site manager Michael Elliott, and the union reps, were strong supporters of the original intention behind the business review meetings: to inform and consult staff in detail about the business strategy and performance of the site. Nevertheless, they had all been frustrated that the information presented had only reached a few members of non-management staff, mainly the four union reps who attended the meetings. To resolve this dilemma, Elliott decided, with the agreement of the union reps, to upgrade the rather ineffective team briefing system into a substantial two-way information and consultation mechanism. In so doing he was following the higher profile given to team briefing within the new division. From now on he would prepare an extended core team brief every month. It would contain around a page on developments in the company and division (prepared by ECD managing director, Keith Parsons), and three pages of detailed information about developments on site under a number of headings: commercial, production, planning, purchasing, distribution, systems, personnel. This would be produced professionally by the site office administrator and copies would be distributed to all staff. Line managers would discuss the team brief at the monthly operations review meetings and then hold briefings with their own staff. Any outstanding questions arising from the briefings would be referred directly to Elliott via the internal electronic mail. He undertook to reply to each question within a week. To enhance communication, Elliott also sent a regular monthly memo to staff reproducing all the questions to him from each team briefing group and all his answers (every member of staff had access to a computer at or near their place of work).

### Pay freeze and renewed changes in the personnel function

Then, in November 1991, only two months after he took up his new job as site manager, Elliott announced formally to all staff that the whole site would be subject to a four-month pay freeze from 15 December to 15 April 1992. Pirelli General – now Pirelli Cables – had made a loss in 1990 for the first time in its history, the recession was biting, production had been cut back, and the rest of the company's factories, whose bargaining date was three months ahead of Aberdare, had been subject to a similar freeze. Against this background and the more general imposition of wage freezes across UK manufacturing industry, the announcement was accepted reluctantly, but without major opposition.

When the other factories emerged from the freeze in January 1992, they were awarded increases of between 4 and 5 per cent for the remaining eight months of the negotiating year. However, ECD managing director Parsons was reluctant to give an equivalent award to the Aberdare workforce. He had two main worries: the continuing losses at Aberdare, and mounting concern over the lack of financial controls exercised on site. However, Elliott argued that these matters were largely beyond the control of the workforce, and that they had done everything they could to cooperate with management (see Chapter 9). In the end, Parsons, Elliott and divisional personnel manager Elaine Jenkins agreed in late January 1992 that the Aberdare staff should be allowed a maximum of 4 per cent.

In the meantime, Aberdare personnel specialist Dick Unwin had left the company to help set up another greenfield site in South Wales. He had enjoyed his job and also felt a strong sense of personal commitment and loyalty to David Regan. The company restructuring in September 1991 changed all that. When Regan left Aberdare to become ECD commercial director and Unwin's job was downgraded from personnel manager to personnel officer (Aberdare was in fact the only site in the new division to retain a management level personnel specialist at all), he decided it was time to go. He was the fourth personnel specialist to leave Aberdare in as many years. The job that was advertised to replace him was a far cry from the original administration manager's job held by David Lane: responsibility for purchasing and accounts had already been lost in 1989; training had reduced in importance; health and safety had now become a divisional (ECD) responsibility. The new personnel officer was no longer to be managed directly by the site manager, but by the new ECD personnel manager, Elaine Jenkins.

Without a personnel manager in post, Aberdare site manager Michael Elliott conducted the 1991–2 wage negotiations on his own – the first time this had been done without David Regan. Elliott assumed that they would follow the procedure established in Regan's day. He told the union reps the total amount he had available – they did not realize that the 4 per cent had only been agreed very reluctantly by Parsons and Jenkins – and expected that they would discuss jointly how to distribute it. But the union reps were annoyed at being offered slightly less than employees in the rest of the company, particularly as they believed they had done everything the management had asked for in terms of flexibility and had cooperated fully from the very beginning – unlike many of their colleagues in other PG factories – with the introduction of total quality management (TQM, see Chapter 9). They put the offer to the workforce and it was decisively rejected. They then returned to negotiate further with Elliott, who twice left the meeting to ring up Parsons to see whether he could offer a little more. It was all to no avail. This also confirmed the reps in their view that, unlike in David Regan's day, the factory was now being run from 250 km away by people who did not understand, and were not committed to, the Aberdare philosophy.

In March 1992 the union reps registered a failure to agree and negotiations moved to the next stage in the procedure: a meeting between Michael Elliott and the full-time regional officer David Evans. Although Elliott had been at the factory since September 1990 (first as operations manager, then as site manager), he had never met Evans before. Relations were cordial, but no agreement was reached. For the first time in the history of the factory, a dispute was referred to ACAS arbitration.

By this time, a new personnel officer, Michael Hawke, was in post. He had previously worked as a personnel officer at Panasonic's television and microwave assembly factory in Cardiff. The Aberdare job offered him his first chance to 'go solo'. After two joint interviews with Michael Elliott and various members of the ECD personnel department, he began work on 9 March 1992. During his interviews, he had been left in no doubt of the two major personnel priorities at the site: the pay structure and the system of skill modules. But he soon found himself faced with a more immediate industrial relations crisis.

Before the pay dispute could reach arbitration, Elliott was summoned in late April to a meeting in Eastleigh with Parsons and Jenkins. They discussed the depressed market for building wires, the negative adjustments that had had to be made to the Aberdare accounts, and the projected loss for the remainder of the year. Against that backdrop Elliott was told that the current staffing levels at Aberdare were untenable. Jobs would have to go. The only question was whether to complete the pay deal and then announce job losses, or to put the pay negotiations on hold, carry out the job cuts, and then implement (and backdate) any pay increase. They decided on the latter approach. At the same meeting Elliott was charged with drawing up a plan to reduce the Aberdare payroll – which now excluded commercial, warehouse and planning staff, who following reorganization were counted as 'divisional' (ECD) employees – by 20 per cent, from 146 to 116.

## The Redundancy Programme

In early May, a short notice signed by Parsons was circulated to all staff at Aberdare. It confirmed the continuing losses sustained by the factory and announced that, in order to reduce costs, the offer of a 4 per cent increase in salary had been withdrawn. Among other cost-saving measures, there would also be an 'unavoidable reduction in staffing'. Consultations would take place, and staff would be informed of details as soon as possible. At the same time Parsons emphasized the company's and division's commitment to Aberdare and pledged full support for the factory in its attempt to become profitable.

Aberdare was not the only part of the new division to suffer at this time. Prices in the mains cables market had fallen substantially in 1991 and planned investment in the factory was withdrawn in the spring of 1992. However, Aberdare staff were the only group in the company not to receive an annual pay award in 1992. Many staff believed they were being punished for not accepting the company's offer. Even worse, rumours began to spread that the site was being prepared for sell-off to another company. During the following weeks rumours spread like wildfire. Predictions of total job losses varied from 15 to 45. On 7 May Elliott informed staff that the details of the redundancies would be announced at two meetings – at 11.00 a.m. and 2.00 p.m. – on 18 May in the presence of divisional managing director, Keith Parsons.

**Table 8.1**  *Proposed job losses, May 1992*

| | Number of staff | |
| --- | --- | --- |
| | 30 April 1992 | Proposed May 1992 |
| Producers | 86 | 70 |
| Maintainers | 25 | 16 |
| Administrators (including systems engineers) | 35 | 30 |

## The announcement of redundancies

Elliott began his presentations on 18 May by announcing that the factory was still making a loss and production was down on the previous year due to the continuing recession. In short, 'if you look at the volume of business, the factory is not viable'. This was underlined by Parsons, who pointed out that the factory could only continue to operate because it was part of a larger division. The division was prepared this year to make good its losses and pay for its central service costs, such as distribution, sales, accounts, research and development. In return, the Aberdare site must put its house in order and make its contribution to stemming the losses.

Elliott then put up an overhead which showed that 30 jobs were to be lost (see Table 8.1). The job losses had been worked out with a small group of line managers on site, but this was the first time that the staff had learned about the details. Elliott announced that the job cuts would save £500,000 per annum on the 'core' staff wages bill, but the losses would be offset to a certain extent by 13 vacancies in the new warehouse – part of the new divisional logistics department and therefore not on the Aberdare payroll. Anyone wishing to transfer to the warehouse was requested to make this known. Volunteers for redundancy were also required. Details of the severance package on offer for each individual could be obtained in confidence, and with no obligation, from the new personnel officer Michael Hawke. If the requisite number of volunteers was not forthcoming, an internal assessment process would be set in motion. This would be done formally, with explicit criteria. A meeting was scheduled with union representatives and full-time officer David Evans to agree a procedure for handling redundancies.

There were only a few questions at the meetings: about the exact timetable for letting people know the outcome (Elliott promised this within two to three weeks); the criteria for the assessment process; whether there would be more redundancies in the future; and how the factory would be able to cope with only 16 maintenance staff (12, plus four senior monthly staff). In response to this last question Elliott announced that they had a plan for the reorganization of maintenance, but that details would have to wait until the redundancies were decided. No one mentioned the pay increase. The two or three staff who were normally the most vocal at meetings were conspicuously and eerily silent.

## The severance deal

Straight after the meeting, two system engineers were informed that they were to be made redundant. This left a systems department of six – the chief systems engineer, four systems engineers, and one systems technician – compared with 13 in 1989 and the first half of 1990. Under such circumstances it was becoming increasingly difficult to justify the post of chief systems engineer within such a small department. (At the end of March 1993 he left the company and was not replaced.)

Personnel officer Hawke informed union representatives of the proposed severance deal for non-management staff on 20 May, just two days after the open meetings. It was virtually identical to the scheme which had applied the previous year to staff in the telecommunication cables factory in South Hampshire. It comprised the small statutory redundancy payment, plus a 'company supplement' based on length of service and age amounting to nearly double the statutory amount. In South Hampshire 300 job losses had been announced, but the scheme, together with early retirements, had resulted in only 15 compulsory redundancies.

For Hawke the last two weeks in May 1992 were extremely hectic. He saw around 20 staff who wanted to find out about their severance payment. Together with Elliott he took part in two negotiating meetings with union reps and the full-time officer about criteria for selection for redundancy. He also held separate meetings with line managers to determine which staff might be made compulsorily redundant and which might be redeployed. In fact, shift managers had already carried out an assessment of their staff before the open meetings in anticipation of the redundancy announcement. However, they had been warned just in time by divisional personnel manager Elaine Jenkins that an industrial tribunal case in another company had recently been lost because of inadequate training of the managers to carry out the redundancy selection. Training – albeit cursory – was hurriedly arranged.

## Criteria for assessment for redundancy

Negotiations with the union on selection criteria started from a management proposal based on the system used in South Hampshire the previous year. There were five main criteria, each with a maximum number of points attached: attendance (12 points), disciplinary record (10 points), attitude (10 points), performance/productivity (10 points), and skills (10 points). Union reps expressed their broad agreement with these criteria, but were concerned that there was no consideration of 'experience'. Elliott and Hawke argued that, in so far as this was relevant, it was included in the 'skills' category. However, they eventually reached a compromise by adding an extra experience element to the skills category, which now had 15 points.

The assessment system was embodied in an agreement on procedures for handling redundancy signed by Elliott, Hawke, GMB full-time officer David Evans and the union branch secretary in early June 1992. Under the scheme, each member of staff would be assessed on each of the five categories and the points would be added up to arrive at a total points score. Every member of staff would have the right to see their line manager to be informed of their score and the breakdown. If compulsory redundancies were to be required, those with the lowest scores would be selected. There would be no appeal, apart from the statutory right to apply to an industrial tribunal that the selection had been unfair.

### The assessment exercise

All producers were assessed by their own shift managers and the manufacturing manager. The assessment of maintainers was carried out by the maintenance manager and his deputy. Assessment of administrators had been done by line managers earlier in the year in connection with divisional reorganization. However, because the required number of administrators applied almost immediately to take voluntary redundancy – one had been left in little doubt that she would be selected if she did not go voluntarily – there was no need to repeat the exercise.

At one stage it looked as if there would have to be five compulsory redundancies among producers and maintainers. So, early in June, the five employees with the lowest points scores were summoned to Hawke's room and told individually, in the presence of their union reps, that they were to be made redundant. However, by a series of reorganizations in management, the creation of some new posts (mainly in the warehouse, where it was clear that staffing requirements had been underestimated), and the transfer of staff from one department to another (again mainly from production to the warehouse), only one member of non-management staff was actually made compulsorily redundant. By the end of July 1992 the target of 30 job losses had been achieved. Twelve staff had left the company, one had already handed in his resignation before the exercise, and 26 managers and staff had transferred to different jobs.

### Job restructuring after the assessment exercise

Apart from the transfer of 17 producers to the warehouse, the most significant outcomes of the restructuring were in production management, shift organization, and maintenance. In production management, the two longest-serving shift managers were moved sideways to different jobs, thus allowing an injection of new blood. Both vacancies were filled from within: one by a shopfloor production operative with previous supervisory experience, the other by a senior maintenance engineer. The redundancy exercise also saw the creation of three new posts of producer/maintainer, one for each shift. These were initially employed in the new PVC compound plant. Here there was no need for full-time maintenance support and Elliott wanted to avoid having to allocate one fully-trained maintainer to the plant on a permanent basis. The introduction of this new hybrid grade of producer/maintainer in the compound plant soon became the prototype for a longer term development programme – beginning in October 1992 – in which around a dozen producers were trained progressively to carry out more of the routine planned maintenance tasks (see Chapter 10 for more details).

The redundancy exercise also had a number of indirect effects. Some highly qualified staff (e.g. in maintenance) left voluntarily, partly because they did not like the way they had been treated over pay and redundancy, partly because they felt the whole atmosphere had changed and they did not see a future for themselves or for the plant. Absenteeism, which was relatively low for the area at around 3.5 per cent in the first half of 1992, fell to under 2 per cent in the six months to the end of 1992. The few staff who had previously been grudging and compliant in their attitude to work and the general Aberdare philosophy 'suddenly' became extremely cooperative and helpful.

The redundancy assessment was the first time that many staff had had their individual performance evaluated (the single-union agreement had envisaged 'regular assessment of performance', but this had never been implemented). They now had clear criteria as to how their performance was to be judged, and many had been pleased at their high points score. They had also seen that managers were not afraid to use the assessments in deciding on job losses. From a mixture of fear of losing their jobs, awareness of the criteria by which their work would be judged, and proof that these would be enforced, the redundancy exercise clearly changed the behaviour of the small number of 'less cooperative' employees.

# Renegotiating the Single-Union Agreement

## *Revising the pay and grading structure*

At the time of the redundancy programme in the summer of 1992, the withdrawal of the 4 per cent pay offer in April still rankled with the Aberdare workforce. However, the shock of the redundancy announcement, the high level of unemployment in their locality (15 per cent in 1992, but nearly twice as high for males seeking full-time manual jobs), and the continuing recession in the cable industry and the country more generally, all meant that the workforce was hardly in the mood for a fight.

Following the completion of the redundancy exercise in July 1992, personnel officer Michael Hawke convened a small management review group, consisting of Elliott and three senior line managers (manufacturing,

**Table 8.2**  *Revised grading structure for non-management staff,*
*1992*

| | |
|---|---|
| Level 3 | Maintainers (automation technicians) |
| Level 2 | POMS system technicians, warehouse team leaders, qualified producer/ maintainers |
| Level 1 | producers, warehouse operators |

maintenance, warehouse), to discuss the type of pay and grading system they would like for the future. The plan was to work out a clear set of proposals, including proposals for a pay settlement for 1991–2, secure approval from divisional management, and then enter negotiations with union representatives over the whole package.

The main outcomes of the internal deliberations were to reaffirm most of the original job definitions, confirm three new positions of 'team leader' for each of three warehouse shifts, and grade warehouse operatives equally with producers on condition that they could be deployed flexibly in the factory if required (most had already acquired production skills before transferring to the warehouse). Under the new scheme there were to be three basic non-management grades – Band A, Band B and Band B plus skill supplement (see Table 8.2). Producers and warehouse operators were to be on the lowest grade, systems technicians (producers with additional systems training on POMS), warehouse team-leaders and qualified producer/ maintainers on the middle grade, and maintainers – all now described as automation technicians – on the top level. The small number of administrators on the Aberdare payroll, who were paid an average of 20 per cent above local rates, were to be taken out of the grading system altogether and their jobs assessed individually. The original ideal of paying all non-management staff the same to emphasize their equal contribution had undoubtedly been distorted and manipulated over the years. It was now to be shelved completely.

## The review of the skill module system

As for skill modules, the management review group decided that all staff would have their maximum 'capped' at the mid-1992 average. For producers and warehouse staff this was five, for maintainers three. There would be no time-served modules in future. For newcomers or those with less than the maximum, one module would be taken every nine months. At the end of this time there would be a standard technical test (as in the past), but now, in addition, reviews of each individual's time-keeping, absenteeism, and attitude (the criteria recently used in the redundancy review). If these were all deemed to be satisfactory, the module would be awarded. If not, it would not be awarded, and line managers would be required to consider the possibility of remedial action and/or disciplinary action for unsatisfactory performance. New warehouse operators would take two warehouse modules, plus two modules in packing and one in assembly, the two production areas closest to the warehouse.

The Aberdare personnel department would now play a central role in monitoring and enforcing the skill module system. It would keep a master copy of all skill modules awarded and currently being taken, and would inform the relevant line manager when a nine-monthly review was imminent. Following the review, the line manager would sign off the module, or identify other action taken, and return a report to the personnel department. Only the personnel officer could authorize the salaries section to pay an additional module (in the past shift managers had sometimes instructed the salaries clerk direct). All staff with six or seven modules would have payment for the additional modules reduced by 25 per cent each year, so that by 1996 no one was being paid for more than five.

As for the new producer/maintainers, they would be recruited on a competitive basis from among experienced producers who had already achieved five modules. At the end of their training they would be expected to carry out most maintenance tasks. In order to reach that level, they would be required to study five modules in the company's Open Learning programme: pneumatics/hydraulics, electrics, electronics, analogue systems, and digital systems. Once they had been signed off (by Bob Walker of Aberdare college and the site maintenance manager) as having completed a particular module, they would be given an extra skill module payment. When they had completed all five, they would be upgraded to become a full producer/maintainer. If they were unable to complete a particular module, or took too long over it, they would lose all their maintenance modules and have to revert to the basic producer grade.

The proposal was worked out in August 1992, and the package presented by Hawke to divisional personnel manager Elaine Jenkins at Aberdare in early September. Jenkins then presented the package to divisional MD Keith Parsons in late September, who gave it his blessing.

## Negotiations with the union

By this time it was only two months before the 1992/3 pay review date, and Parsons, Jenkins and Hawke decided to put a three-part package to union representatives: the revised pay and skill module system, a pay offer for 1991/2, and one for 1992/3. Just before these were presented to union representatives, the September sales and production figures for Aberdare were announced. They made it abundantly clear that the factory was still a long way from making a 'positive contribution' to the division.

Against this background, and under pressure from Parsons and Jenkins, there was to be no increase at all on basic rates for 1991/2. However, in recognition of staff cooperation with TQM and the fact that many of Aberdare's problems were beyond the control of employees, all non-management staff were to be offered a one-off non-consolidated payment of

£150 – between 1 and 1.5 per cent of gross earnings. Finally, there was to be an across-the-board 4.3 per cent increase on basic rates from December 1992, and a small increase on shift allowance. Skill module payments would remain at £300 each.

Before these proposals were made public, they were put to union representatives by Hawke on 16 October. Hawke had also arranged for David Evans, full-time union officer, to be present, both because the proposals represented a variation on the single-union agreement and because the four union reps were relatively inexperienced. Only one of the reps had been present at the negotiations in March/April, three of the original four having resigned after the redundancy: one had become the new shift manager (occasioning a number of unprintable remarks about the role and career strategies of union representatives!); the other two had simply had enough, exhausted after having had to spend so much time and energy around the time of the redundancy. This had included having to be present at the often emotional and tearful meetings when individuals were told they were to be made redundant.

After Hawke had outlined the package, the new union reps said that they did not disagree with the broad structure, but were not happy with some of the details. After holding a branch meeting, they met with Hawke again and requested a larger lump sum for 1991/2, an increase to seven in the number of skill modules for producers, and the immediate award of two modules to warehouse operators in recognition that they were already working flexibly across the warehouse.

Hawke immediately consulted Elaine Jenkins. The negotiations were clearly taking on an 'old-style' character, although this was not surprising given the events of the previous nine months. Nevertheless, since it appeared that the company would be able to achieve nearly all its objectives, Jenkins and Hawke agreed that they would give union representatives some concession, however small, to take back to their members. They therefore decided, with Parsons's approval, to increase the 1991/2 payment by £50 to £200. This was presented to the third negotiating meeting on 30 October and put to a ballot of the whole workforce on 2 November (the voting box was located on the reception counter at the factory entrance). On 4 November 1992, the day of the USA Presidential election, the votes were counted (94 per cent turn out) and the result announced:

101 issued
 79 in favour
 16 against
  6 not returned

The acceptance was overwhelming. It was greeted with a collective sigh of relief on all sides. Most staff had been extremely fearful of having the offer withdrawn again (as in April) if it had been turned down. They were also acutely aware of the continued lack of profitability in the factory. They

**Table 8.3    *Union density at Aberdare, 1990 and 1992 (%)***

|  | 1990 | 1992 |
|---|---|---|
| Shopfloor (producers/maintainers) | 100 | 99 |
| Office (administrators) | 53 | 76 |
| Overall | 92 | 95 |

were particularly thankful that they still had a job. The new system came into operation on 1 December 1992.

# Changing Perceptions of the Union Role

Two interview surveys of union members (see Technical Appendix for details) give an interesting insight into their views of the changing role of the union at Aberdare between 1990 and 1992. The main continuity between the two surveys was the remarkably high level of union density, particularly among shopfloor manual employees (see Table 8.3). However, there had been an increase in union density among the comparatively small number of administrators, as company restructuring and the continuing recession intensified the threat of redundancy and job loss. The reasons for the continuing 'status' difference in union density levels are relatively easy to explain. The union branch committee of four members was dominated from 1987 to 1992 by shopfloor employees, often with substantial past experience of trade union office. Virtually all the shopfloor employees at Aberdare had been union members in their previous employment, many had a strong commitment to the ideas and traditions of trade unionism, and the rest joined when asked 'because everyone else was a member'. In contrast, most of the office employees joined the company in 1988 after the main shopfloor recruitment drives were over, many had not previously been members of a union, and when they started employment at Aberdare there was no union representative in the office to recruit them (there was no provision for separate representation of shopfloor and office staff in the branch committee). Of the eight administrators who were not union members in 1990, five said they had not been approached to join the union, and the others either did not approve of union membership or did not think it was necessary in the 'open' Aberdare environment. Interestingly, although around three-quarters of administrators were female, the small number of male administrators accounted for half the total number of non-union members on site.

If we look at union members' perceptions of the most important aspects of the union role, we find that pay negotiations were mentioned by only around one-seventh of union members surveyed in mid-1990 (see Table 8.4). This clearly reflected the low actual and perceived level of union influence in the

**Table 8.4** *Members' perceptions of most important aspects of union role at Aberdare, 1990 (n = 95)*

| | |
|---|---|
| Protection/sorting out grievances | 69 |
| Support in cases of accidents and health and safety | 32 |
| Negotiating pay and conditions | 16 |
| Legal support | 13 |
| Others | 5 |
| No answer | 14 |

The responses add up to more than 95 as some respondents gave more than one answer. The most popular response – protection – could be said to include legal support and accidents, too. However, these have been listed separately where they were mentioned separately from protection and grievances.

context of the 'new style' of negotiations in 1988 and 1989. The main role of the union at this time was seen as one of 'insurance', whether in terms of grievances, accidents, or other problems (only two employees mentioned discipline as a potential problem). Overall, the union representatives were felt to be doing as well as they could under the circumstances of a 'very tight' agreement in which management very much held sway.

When a similar question was posed in 1992 ('what is the main role of the union on site?'), negotiations over pay and conditions were still mentioned by one-eighth of respondents, but a quarter volunteered that the union was now totally powerless to influence management on pay. It should be remembered that the interviews took place in July and August 1992 when the imposed pay freeze was still in operation, the pay offer for 1992 had been unilaterally withdrawn, and the redundancy exercise had just been completed. The most important union role now, according to union members, was protection in case of accidents, and health and safety (a quarter of all respondents). In fact there had not been many accidents over the five years since opening, but where there had, union members had been able to take advantage of professional advice as well as organizational and legal support.

There was no unanimity among union members about how to make the union more effective in future. The vast majority certainly did not wish to return to the industrial relations of the 1970s. As one of the branch committee members said: 'all that nonsense about going out of the gate [on strike] is over, nobody wants that'. On the other hand they did not want a return to the days when managers could simply do what they liked. In this context, quite a few members remarked unprompted that, relative to other places they had worked in the South Wales valleys (and in comparison with the old Pirelli factory), pay, conditions and health and safety were extremely good at Aberdare. If there was some kind of consensus, it was voiced by one of the new members of the branch committee, with substantial experience as a lay union officer in his previous employment:

What we want is more of a works committee than a union, we want involvement. [Over the past two years] the union tended to bring things in from outside and operate secretively. What we want is a works committee that collates things from the shopfloor, puts it forward to management, comes and reports back, and puts things on the notice board for all to see and read. So far the union here has gone into discussions with management with a blank sheet of paper. We need some demands and ideas. If the management is sensible, they will listen, take some on board, reject others, say they will investigate others. We don't expect everything to be agreed. We do expect to be listened to and given answers.

There were two great contrasts in the experience of most non-management employees at Aberdare in 1992. The first was between the optimistic expectations with which they had come to the factory in 1987 and the harsh realities of 1992. The second was between the feeling of powerlessness in areas such as pay and job cuts, and the feeling of empowerment through self-supervision and involvement in total quality management (TQM) and continuous improvement groups. TQM is the theme of the next chapter.

## Summary

This chapter has traced the design and operation of the Aberdare single-union agreement from 1985 to 1992. The first part looked at the initial discussions about union recognition and the consultations in South Wales about the procedures for selecting the single union. The company eventually followed the procedures recommended by the Wales TUC and, after a beauty contest involving five unions, selected the GMB as the most accommodating, moderate and forward-looking union in the region.

After conventional negotiations in 1988, union representatives agreed in 1989 and 1990 to a new form of pay negotiations. These resulted in increases slightly above the rate of inflation, but favoured those groups of staff – above all maintainers – for whom there was the strongest external labour market competition. Business review meetings started in mid-1989 and proved a highly successful if numerically limited forum for consultation about the trading position and performance of the factory. In the wake of company reorganization, these meetings were superseded by a more systematic and upgraded form of monthly team briefing.

From the end of 1991 the Aberdare site experienced a year of turbulent change in the operation of the single-union agreement. A pay freeze was followed by a redundancy exercise and then a renegotiation of the agreement. This involved a restructuring of the grading system more in line with external labour market pressures and a further restriction on the number of skill modules that could be achieved. Union membership remained extremely high throughout the 1987–92 period, but the poor commercial position of the company and the site meant that, by 1992, most union members felt almost completely powerless to do anything collectively or individually about pay and jobs.

# Questions for Discussion

1 Analyse the reasons why Pirelli decided to opt for a single-union agreement for its Aberdare plant.

2 Discuss the procedures and mechanisms used to select the single union.

3 Which union would you have selected, and why?

4 What were the strengths and weaknesses of the new style of pay negotiations conducted between 1988 and 1990?

5 Evaluate the operation of the agreement until the end of 1992 from the perspective of (a) site management (b) the union members on site and (c) the GMB.

# 9

# Managing Quality and Continuous Improvement

## Thematic Focus: Total Quality Management

We saw in Chapter 6 that 'quality' is one of the key features which distinguishes human resource management from traditional personnel management. Indeed, it has been argued that total quality management – the focus of this chapter – is a 'fundamental and necessary element within the HRM project' (Sewell and Wilkinson, 1992: 101).

The idea of TQM has its roots in the USA, but it was first taken up systematically in post-war Japan. It grew out of the application of rigorous quantitative performance techniques – such as statistical process control (SPC) – to manufacturing processes (see Hill 1991: 399). It was subsequently developed into a fully-fledged theory of management applicable to all types of organization – public or private, manufacturing or service, large, medium or small.

For its leading proponents such as Edward Deming (1982, 1986), TQM means shifting the emphasis from inspection for quality after the event – usually via a separate quality control department – to preventing defects or poor service through the integration of quality management into all aspects of an organization's operation. The holy grail of TQM in commercial organizations is to create a virtuous circle in which the elimination of waste ('zero defects', 'right first time') leads to improved quality, lower costs, greater competitiveness, increased market share, and ultimately long-term profit maximization (short-term profit maximization is alien to most theories of quality management; see Hill, 1991: 403). Since 85 per cent of all quality failures are held to be the result of inadequate management systems (Ishikawa, 1985), there is a strong emphasis in the TQM literature on the central importance of management in establishing quality systems.

The leading advocates of TQM in the UK came initially from production and operations management (see Oakland, 1989). An increased concern with quality was also fuelled by pressures on companies to conform with the quality rules and procedures of the British Standards Institute, as indicated by the award of its quality standard BS5750 (see Wilkinson et al., 1992: 1–4). By the 1990s, quality management had become a 'general management concern' (Wilkinson et al., 1992: 2) for organizations in all sectors and industries. However, implementation in the UK has so far been 'uneven and sparse' (Geary, 1994: 643–4), although it has been almost universally adopted by large foreign-owned multinationals.

As with HRM, there is no consensus in the research literature about how to define TQM. Like HRM, too, some definitions stress 'harder' production- and performance-oriented goals, while others point to the 'softer' elements of TQM such as employee commitment, teamwork, involvement, flexibility and training. There is now general agreement, however, that TQM involves not just the use of a single management technique – whether SPC or quality circles – but the linking of a range of techniques and procedures into a 'holistic system of management' (Hill, 1991: 399).

Building on the work of Hill (1991: 400–1), it is possible to identify seven main principles underlying TQM:

1   The product or service must meet the requirements of *external customers*. This is the principle of 'fitness for use' (Juran, 1989). It implies both quality of design – ensuring a product or service is designed and specified to meet the needs of the customer/client – and quality of realization – ensuring it actually does what it is designed to do.

2   A product or service must also meet the requirements of *internal customers*. TQM requires staff to see each other as customers and to meet each other's requirements for a high quality, and ideally fault free, product or service at each stage in the process.

3   TQM requires routine *assessment* of the product or service using appropriate *performance measures*, with corrective action to be taken if necessary.

4   *Executive level coordination* is required to ensure TQM is applied consistently throughout the organization. This usually takes the form of a steering committee of senior managers, which establishes corporate policies and procedures and monitors outcomes.

5   TQM requires *employee involvement* in decision-making, often via new organizational arrangements. It also requires champions at lower levels within the organization, often via mechanisms such as project teams or quality improvement groups. Unlike quality circles in the UK in the 1980s, which were mainly voluntary and involved only around 10 per cent of employees, TQM is based on the idea that all employees have relevant knowledge about their particular work area and all can, and should, contribute to improved performance.

6   TQM requires an appropriate *organizational culture* if everyone is to endorse the total quality objectives and follow TQM procedures. This culture needs to include high-trust relationships and a shared sense of common objectives.

7   Finally, while it is managers who have the initial responsibility to create the environment and systems which encourage a TQM culture, total quality results ultimately from *a practical commitment to continuous improvement* from every employee at every level in the organization.

As so often with models and ideals, whether called CIM, HRM or TQM, the ultimate result depends on the balance struck between various components of the system and the way in which they are implemented. Recent evidence suggests that initial hopes and enthusiasms can often give way to disillusion, leading ultimately to a failure to sustain the initiative beyond a few initial experiments (see Wilkinson et al., 1992: 14–17). There appear to be four main reasons for this.

First, many TQM initiatives have been confined to particular activities and departments and not integrated into wider corporate and commercial strategies. Second, TQM can become the subject of conflict between different management groups, with some groups – in manufacturing environments, particularly those in production management who tend to gain most from the initiatives – imposing themselves on others and thereby generating resistance. Third, senior managers have often regarded TQM as an exclusively managerial initiative, making little attempt to establish a positive and cooperative working climate or to gain positive employee – and, where applicable, union – support for the programme. Finally, there is a potential conflict (as well as complementarity) between TQM and employee involvement. While the language and philosophy of TQM is very much about empowering employees, there is also a strong emphasis on increasing management control by tighter surveillance of staff and monitoring of performance (see on this Sewell and Wilkinson, 1992; Garrahan and Stewart, 1992).

In summary, the ideal of TQM involves the achievement of certain quantifiable objectives, such as the elimination of waste, zero defects, right first time. It also implies the use of certain methodologies and techniques, such as SPC, JIT scheduling, quality circles, and the use of various charts and boards to make objectives and improvements explicit (sometimes dubbed 'management by display'). At its heart, however, is the idea of continuous improvement (in Japanese, *kaizen*) of all aspects of an organization's activity.

---

# Reorganization and Changing Multinational Objectives

In 1990, Pirelli General reported an overall trading loss for the first time in its history. In early 1991, on the initiative of sector management in Milan, the company launched a TQM programme. The two events were not totally unconnected. Managing director Felipe Martinez admitted as much in a circular distributed to all employees on 12 April 1991, headed Total Quality Management. According to Martinez the company had not been performing well over the previous few years. Despite a major programme of investment and reduction in corporate overheads, profitability had continued to decline. Employee morale and customer satisfaction were also much lower than they should be. He concluded: 'senior management have been considering the options and have decided that the introduction of TQM will provide the necessary stimulus to address these outstanding problems'.

By the time Martinez had issued the circular, senior and middle managers had already set the framework for the introduction of TQM. A team of leading management consultants had been engaged to help in implementation; a corporate TQM manager had been appointed to lead the initiative; all senior company executives had attended a TQM training course during which they had helped draw up a company mission statement; a corporate quality council – chaired by Martinez and composed of the four divisional managers and heads of the main corporate departments – had been set up to coordinate the implementation of the programme; 60 senior managers had met to identify the areas in which quality improvements were most urgently needed in each division; and surveys had been carried out by management consultants to establish how customers and employees regarded the company.

The next phase of the programme involved setting up 'fast track' project teams in each division to tackle the areas or problems identified by senior managers as in need of urgent quality improvement. Initially, only a small number of staff would participate, but within the following seven months all 1800 company employees would be asked to participate in a TQM awareness course. By starting with a small group of 'fast track' projects, the company intended to achieve clearly defined improvements in priority areas during the early stages of the programme. It was hoped this would show that TQM was not just the latest 'flavour of the month', but a serious attempt at achievable quality improvement.

The initial reaction of managers at Aberdare was one of positive commitment. In contrast, one long-serving senior executive at corporate level adopted a more cynical, if still broadly supportive approach:

> There is nothing special about TQM . . . What TQM has done is to take an old idea and dress it up in a lot of razzamatazz, and also to make all the consultants rich with their packages and mission statements and ticks in boxes. . . . It is simply good management and we should not have a need for it in a company worth its salt. But we don't live in a perfect world, so therefore I think it is a wonderful thing if you can get the collective whole of Pirelli to believe it's a good thing. I don't know a better way, and therefore TQM could well be the solution to improving across the whole . . . But it will be like a snowball and will gather momentum after a very slow start.

There was indeed a very slow start. The first nine months of the programme were overshadowed by a series of events which, taken together, gave the company a shake-up which touched every one of its 1800 employees. It is not too much to say that 1991 marked the end of an era for Pirelli General and the beginning of a new period of almost continuous change.

## Changing sector objectives

On 15 March 1991, the old Southampton general wiring factory finally closed after 77 years of operation. The building had also housed the corporate headquarters, which transferred in a thinned down form to a new building on the company's Eastleigh site, eight km north of Southampton. Apart from the Aberdare factory and the submarine cable plant in Southampton, all of Pirelli's UK cable-making operations were now located on two sites within 2 km of each other. Meanwhile, at group level in Milan, a new world-wide cable sector company, Pirelli Cavi (Pirelli Cables), had been created. This brought together the group's world-wide cable interests in one wholly-owned subsidiary (until then the Italian cable factories had been part of a separate Italian holding company; see Chapter 1). At the same time, Pirelli General was renamed Pirelli Cables to give it a clearer product and market profile and help it emerge from under the shadow of the more glamorous tyre sector.

One of the first acts of the new senior management team of the Pirelli group was to set up a separate European organization based in Milan. Its function was to coordinate the activities of the Pirelli cable companies in Italy, the UK, Spain and France in anticipation of the creation of the new European Single Market on 1 January 1993. The new European organization had its own managing director with direct line responsibility for the managing directors of the four national cable companies.

As if these changes were not enough, Pirelli Cables (UK) announced publicly on 15 May 1991 – the workforce had been told two days earlier – that it had purchased the cable products division of Standard Telephones and Cable (STC) from its current owners, Northern Telecom. The STC division was based in two main locations – Newport in South Wales, about 60 km from Aberdare, and Harlow, in Essex, east of London. With annual sales of over £80 million and around 750 employees, it was over twice the size of Pirelli's equivalent operation at Bishopstoke. At a press conference, the group's cable sector general manager Guido Angeli explained the reasons for the takeover, saying that the merging of the two divisions would strengthen the competitiveness of the company in the market-place and form 'an integral part of our European operations' (*Southampton Evening Echo*, 15 May 1991). In a subsequent statement to all staff, Felipe Martinez announced that Graham Phillips – previously general manager of STC's cable products division – would be in charge of the company's new unified UK telecommunication cables division.

The effect of the merger was soon felt. On 2 August 1991 around 300 redundancies were announced at the telecom cables factory at Bishopstoke. The announcement sent shockwaves through a workforce accustomed to the caring paternalistic culture fostered over many decades. The shock was reinforced by the criteria for selection for redundancy, which included the 'attitude' of employees and their attendance record. Although only around 15 staff were eventually made compulsorily redundant (the rest took voluntary severance or were redeployed to other sites), the whole episode signalled the emergence of a management style and approach more attuned to the harsher realities of the 1990s.

In the midst of all these turbulent changes, in June 1991, cable sector managers in Milan convened a meeting of the site managers of all the group's general wiring factories world-wide. It was to be held at Aberdare. It was attended by the group's cable sector general manager, the European cable sector managing director, and the head of the sector technical department, as well as representatives from most of Pirelli's affiliate national companies. David Regan saw the location of the conference at Aberdare as a major challenge. He knew that the Aberdare experiment was regarded by many within the group and the sector as a failure. Here was an opportunity to demonstrate that Aberdare really was the factory of the future, even if not everything had yet been achieved. Knowing that the operation would be scrutinized by people with long experience of cable-making, he decided to go for broke. A few weeks in advance he announced to all staff that, by the time of the conference, production would run 'without paper'. This meant, above all, the end of the paper-based manufacturing orders (MOs) which were currently attached to each drum and made a nonsense of any claim to be a computer-integrated operation. This had an important cathartic effect on the use of POMS, as we shall see.

Of greater longer-term import for the cable sector as whole, however, was the discussion at the conference of sector objectives for the national cable companies. Return on equity, measured by agreed annual profit targets, was the most important criterion by which national companies and each individual business unit were currently assessed. However, as we saw in Chapter 1, the traditional UK Pirelli culture, reinforced by sector managers in Milan, also placed a strong emphasis on high levels of engineering excellence (the 'Rolls-Royce cable'). It was also characterized by an almost obsessive focus on output and technical and production efficiency. This culture exerted an enormously strong pressure on national company executives and divisional managers, so that production efficiency measures – such as line speeds, machine utilization, output per employee hour – came to be regarded by middle and junior managers as equally (and in some cases, more) important criteria in measuring success.

However, the world-wide decline in competitiveness and profitability of many of the company's production facilities in the late 1980s had led to a rethink at sector and group level. The TQM initiative was one of the outcomes. Now, at the general wiring sector conference in June 1991 in Aberdare, the reorientation was made explicit. Technical and production criteria were no longer to be uppermost, rather profit and loss and the ability to meet customer and market needs. Henceforth, sector and European management would be much more interested in the ability of national companies to reduce operating costs and sell the products customers wanted at more competitive prices, thus generating an improved return on capital. One of the main methods of achieving this would be to gear production to meeting customer requests first time (the so-called 'hit rate' criterion). Production should be geared to the needs of the market and the customer, and not the other way round. Or, to adapt Peter Drucker's famous distinction, the focus was not to be on efficiency alone – doing things right –

but on effectiveness, that is, making sure you are doing the right things. We shall see below that this top-level reorientation gave impetus to a process which was already in train at Aberdare.

# Divisional Reorganization in Pirelli Cables

### The creation of the Energy Cables Division

In September 1991 Martinez announced a change in the organization structure of the UK company which reinforced the group and corporate changes made earlier in the year. The existing structure, with strong corporate departments and four semi-autonomous divisions (telecommunication, power, industrial, and general wiring cables), was to be replaced by a new structure with a leaner corporate HQ and two enlarged divisions encompassing all the current operating sites. The two new divisions were to be organized around the main distinctive product markets and customer bases of the company: telecommunication cables and energy cables. The creation of the telecom cables division was a logical outcome of the takeover of STC's activity in this area and was generally expected. The creation of a new energy cables division (ECD), spanning everything from high voltage power cables (supertension, mains and submarine) to low voltage building wires, had not been expected. However, the savings in overheads and the broad commercial logic of the move were plausible and, for many in the company, long overdue.

Prior to the reorganization, customers wishing to buy a mixture of building wires and industrial cables had had to deal with two separate divisional sales departments and two separate sets of order forms, invoices and deliveries. Following reorganization, a unified commercial (sales and marketing) structure was created to cover the whole of the new energy cables division. The result was that the company was now able to present just 'one face' to the customer rather than two or, exceptionally, three. (The basic organizational principle was similar to the one adopted by British Telecom one year previously. In 1990, BT had abolished its geographically-based districts and created three new 'customer-facing' divisions (personal communications, business communications, and world-wide networks) so that private, business and multinational companies could satisfy all their requirements – customer accounts, sales, faults, complaints, new services, new products – by reference to one unified organization.)

To underline the need for a greater commercial influence on production, ECD managing director Keith Parsons decided to establish a new logistics function within the new division. As a result direct line responsibility for production planning, stock control, purchasing (of raw materials), warehousing and distribution was taken out of the hands of Aberdare managers and transferred to a new divisional logistics manager based at Eastleigh. Martinez had introduced a similar change when he was managing director in

Spain, which had resulted in a greater gearing of production to customer requirements rather than the needs of production.

To reinforce this change in emphasis, Parsons outlined his strategy for the future of the division in two house journals, the company-wide *TQM Review* (September 1991) and the new Energy Cables Division newsletter, *Cable Talk* (February 1992). Like Martinez before him, he began by stressing the poor performance of the company over recent years. He also admitted that the traditional Pirelli General management style had been at fault in not encouraging employees to be actively involved in improving performance. In both articles he stressed the centrality of TQM as the vehicle 'to fundamentally change the way our division operates'. He identified 15 'fast track' projects in the new division, some of which had already started in the summer of 1991, others which needed revitalizing after the upheavals of wider restructuring. The next stage, following the completion of these time- and target-limited fast track projects, was to set up quality improvement groups (QIGs) of between four and eight employees to carry out process and quality improvements within their own designated work areas. He continued:

> The major task for the division and the managers of the division is to push the decision making and the authority for the decision making down through the organization to a level of active involvement, so that the people who are the experts in that particular area – whether it be making cable, repairing a machine, supplying an invoice, or progressing a customer order – can contribute to the solution we need to make the company more efficient and more effective.

Parsons concluded by summarizing what he saw as the four basic aims of TQM (two 'hard' and two 'soft'):

- market focus
- lowest cost production
- employee involvement
- positive management approach to employee contributions

TQM was clearly to be the guiding principle of the new Energy Cables Division.

### Reorganization at Aberdare and the departure of David Regan

The restructuring of the company had an immediate impact on the management structure at factory level. At Aberdare, the three senior business and commercial management positions – divisional accountant, commercial manager and marketing manager – were abolished. David Regan himself, as divisional manager, had previously been in charge of the whole of the business operations connected with the factory and had been profit-accountable for its overall performance. He now lost direct responsibility for accounts, sales, marketing, purchasing, overall production planning, warehousing, dispatch and distribution, and was left simply with daily production planning, production itself, maintenance and, provisionally, systems. Effectively, the unit was now responsible for a production facility, not a commercial business operation.

The performance of each factory site was now to be measured by its 'contribution' to the overall profit and loss of the wider division. The contribution would be calculated in terms of the difference between the selling price of the goods they produced (now the responsibility of the divisional commercial director) and the cost of manufacturing them (materials, labour, fixed overheads, depreciation). Excluded from this calculation were 'logistics', together with the site's contribution to wider company and group overheads. The primary task of the new Aberdare 'site manager', as the position was now to be called, was to concentrate on producing to customer requirements at the lowest possible cost. It was the task of the divisional commercial department to determine the best way to make a profit in the market-place.

Regan fully supported the commercial reasons for the reorganization and the cost advantages that could be gained from abolishing the duplication of activities on the commercial and marketing front. He was, though, fearful of the impact of the new larger division on the technical and human resource innovations at Aberdare. He was also definitely not interested in staying on in the much-reduced site manager's job. He was keen to be appointed as the new ECD managing director. However, as we have seen, this position went to Keith Parsons, who had been divisional manager of the profitable power cables division at the time of reorganization and was held by Martinez to be the best manager available for the post. Regan was offered, and accepted, the job of ECD commercial director with direct responsibility to Parsons. He was replaced as Aberdare site manager by Michael Elliott.

# Total Quality Management and Culture Change

### A profile of Michael Elliott

Elliott had begun his working life as a trainee technician apprentice, and joined Pirelli General in 1985 as maintenance manager for the Southampton site. As the Southampton factory was gradually run down, he was put in charge of the transfer of the plant and machinery to its new site at Bishopstoke. However, his most important career move to date had come in September 1990 when he was appointed operations manager at Aberdare at the age of 42. He had always wanted to go into production management; the only drawback was that his family would not be able to move to South Wales, at least in the short term, due to job and school commitments. He resigned himself to regular 500 km round trips and the inevitable hotel stays during the week.

His first months at Aberdare were spent talking to his fellow managers and working on the shopfloor alongside producers and maintainers. At the end of this period he had formed a clear view of his role and the priorities for the plant:

> The image of Aberdare was that the plant wasn't any good, that the systems area was a disaster, with everyone in the systems department leaving [five had left between June and September 1990]. But I tried to approach things with an open mind. I must admit I was surprised to see the amount of paper that was flying round, it didn't seem to be right . . . What I can say is that the attitude of people here is superb, the commitment, everybody wants to win. Every operator you talk to has five or six ideas as to how the machines could work better, and with three shifts that's well over 300 ideas. It's a bit frustrating that nothing systematic has been done about it . . . I am coming from the outside and see the potential, but I feel we aren't communicating properly and tend to run around in circles. Anyway, I felt we needed to select the top 10 of these 300 ideas and tell everybody what we were doing and why. All I need to do is to light the blue touch paper. My job is to act as a catalyst for things that are waiting to happen. (Interview, 20 November 1990)

And this is what Elliott did, first as operations manager working for Regan, then, from September 1991, as site manager. His initial contribution was overseeing the introduction of Total Quality Management.

## Quality management at Aberdare prior to TQM

A basic level of quality control is a legal requirement for any company producing electrical products. Responsibility for setting and monitoring minimum quality standards in the UK cable industry is vested in the British Approvals Service for Cables, BASEC. To achieve BASEC approval, and with it the award of the quality standard BS5750 (now BS EN ISO 9000), companies must have formally stated policy objectives for product quality, accessible documentation laying down the standards required for each machine process, and a trained workforce which is knowledgeable about these standards. Only if all these requirements are met after an independent inspection can cable manufacturers emboss their cables with the acronym BASEC and issue 'certificates of conformity' with British standards, which many companies – particularly large utilities and overseas customers – demand as a prior condition of purchase. Once approval is obtained, BASEC officials make regular monitoring visits to factories and take away samples for testing. Their ultimate sanction is withdrawl of approval.

When the organization structure for the Aberdare factory was agreed in 1987 (see Chapter 6), no provision was made for a special department to deal with quality. Regan had argued from the start that quality assurance should be built into the CIM system, and that quality management was the responsibility of all members of staff and should not be hived off into a separate department. In the event 'residual' quality assurance duties were assigned to a site technical manager, Peter West, who was a 'temporary' secondee from the corporate technical department.

By early 1989, however, it was becoming clear that the amount of preparatory work needed to gain BASEC approval for Aberdare (the

Southampton factory had already gained approval) was enormous. After many requests West was given a small team of staff by the corporate technical department to carry out this task and they worked extraordinarily long hours to ensure that all the necessary documentation was completed and production staff carried out the required tests. It was not until mid-1990 that Regan and his operations manager Martin Dawson finally recognized the inevitable and agreed to establish two permanent quality assurance posts on the Aberdare payroll: a QA manager with overall responsibility for quality management in the factory, and a QA administrator to carry out the routine work associated with BASEC approval and certificates of conformity.

Most of the product quality problems at Aberdare in the first three years of operation came in fact from lack of reliability in plant and machine processes (see Chapter 5). These were, of course, the direct responsibility of production and maintenance managers and their staff. Indeed, much of the time of maintenance staff between 1988 and early 1990 was taken up with attempts to bring various parts of the plant up to the required reliability level. Most of this work, though, was *ad hoc* and reactive, with the result that many problems were simply patched up rather than systematically examined and engineered out. The outcome was scrap levels of 8 per cent of total output produced.

In 1990, as we have seen, there was immense pressure on the factory management to meet its production output target. Any intervention which involved stopping production, such as taking particular machines out of circulation and giving them a thorough overhaul, met with hostility from production management and staff. Nevertheless, Dawson and manufacturing manager David Thomas agreed in mid-1990 that something fundamental needed to be done to deal with recurring quality problems on the machines that were continuing to produce high levels of scrap (now around 6 per cent). They therefore decided to set up a small 'development team' to work on the most intractable machine problems. It was chaired by the maintenance manager and involved two senior members of the Aberdare technical staff – one specialist automation engineer and one production engineer – together with part-time help from three newly-appointed 'process technicians'.[1]

While the development team did some useful work between mid-1990 and mid-1991 on specific machine problems in metallurgy and extrusion, its function was gradually overtaken, then eclipsed, by the introduction of TQM. This was the vehicle through which Elliott began to focus the effort to optimize the performance of the plant and to generate continuous quality improvements in all aspects of its work.

---

[1] These posts, one on each shift, were created to work on special machine improvement projects and give general production support to the shift managers. The three technicians were recruited from within the production and maintenance workforce and promoted from Grade A to Grade B. They were required to do a two-year Higher National Certificate day release course in production engineering. They were to become important links between the TQM programme and the workforce (see below).

## *Phase 1 of TQM at Aberdare: fast track projects*

The immediate reaction of Aberdare managers and staff to the launch of the company's TQM programme was to say 'we are doing it already'. In some ways they were right. As one noted: 'We've had TQM from day 1, there is no need for a culture change'. This was in direct contrast to the company's factories in Hampshire, where manual workers were initially strongly resistant to TQM and participation in problem-solving teams. And yet, while the general site culture at Aberdare was certainly conducive to TQM, it was, as one manager later remarked, 'TQM without measurement'. It was certainly without systematic objectives, clear priorities and explicit procedures. This was the kind of sharper focus provided by the company's TQM programme.

As we have seen, the first phase of the programme required senior management at each site to identify two or three priority areas for improvement where specific quantifiable and achievable targets could be set. At Aberdare, three such fast track projects were identified (see Table 9.1) by Elliott and manufacturing manager David Thomas in consultation with other managers. All were geared directly to the site's new prime objectives of achieving maximum customer service and lowest cost production.

**Table 9.1** *The three fast track TQM projects at Aberdare, 1991*

| | |
|---|---|
| *Project 1* | *Work-in-Progress* |
| Objective: | To reduce product lead time (measured by the average time taken to produce cable from start to finish) and the value of work-in-progress |
| Members: | Stuart Wood, systems engineer (team leader); Graham Hall, planning coordinator; Michael Bevan, Michael Pace and Andy Ludlow, producers (the first two had recently been appointed 'systems technicians' with special responsibility on their shift for POMS implementation) |
| *Project 2* | *Scrap* |
| Objective: | To reduce the overall percentage of scrap and improve the handling and control of defective material |
| Members: | Barry Pascoe, process technician (team leader); David Norton, Alan Dale and Terry Doyle, shift managers; Barry Miles and Mark Thomas, process technicians; and Mark Phillips, Ted Johnson and Stephen Evans, producers |
| *Project 3* | *Finished Goods Stock* |
| Objective: | Reduce finished goods stock while simultaneously maintaining stock availability for the customer |
| Members: | Graham Hall, planning coordinator (team leader); Mike Poulton, shift manager/temporary warehouse manager; Eddie Bevan, divisional buyer. |

With the upheavals following the acquisition of STC's cable products division and the subsequent restructuring of the company, the three projects did not really get down to serious work until October 1991. Nevertheless, by the spring of 1992 they were already achieving major improvements, and in March the 'scrap' project team won one of the first two quarterly TQM awards in the new division.

The 'scrap' team was chosen by Elliott and Thomas, and deliberately involved a mix of staff from each shift with different kinds of knowledge and responsibility: three shift managers, three process technicians, and three producers. The shift managers were formally responsible for final product quality and were the only people with the power to declare whether a particular project should be scrapped or not. The process technicians were ex-producers and ex-maintainers selected because of their technical knowledge and problem-solving abilities. Finally, the producers – one from each shift to ensure that the ideas from the project were disseminated – were the basic grade staff who were responsible under self-supervision for the day-to-day manufacture of the product. In line with the non-hierarchical Aberdare (and to a lesser extent, TQM) philosophy, it was not a shift manager but a process technician who was made team leader. (Just over two years later he was promoted to shift manager.)

The problem facing the 'scrap' project team was easily quantifiable in terms of the cost of raw materials and lost sales. In 1989, at the height of the machine problems, scrap levels had touched 8 per cent of total production, and they were still averaging 5 per cent in the autumn of 1991. Working under corporate TQM guidelines, the project team started by establishing the overall objectives, preliminary goals, and detailed timescales for the project. They were also required to identify basic assumptions, resource constraints, and project boundaries. At the first meeting, they agreed that the problem of scrap started in the metallurgy section with poor quality strand and short or mismatched lengths. This obviously had knock-on effects and costs for the rest of the manufacturing process. They therefore decided to concentrate their initial effort on identifying the causes of problems in metallurgy. With the help of producers working in the area, it only took them a few weeks to come up with a number of innovations and machine modifications, which they gradually implemented. Within just three months they had reduced scrap levels to 4.3 per cent.

They then moved on to look at problems with one particular extrusion line. After investigations they decided to try out a new mix of plastic insulation material, which soon led to further reductions in the amount of scrap. In March – the high point of their project – total improvements across all sections saw scrap levels down to under 3 per cent. They had achieved their target, and the group disbanded. However, to keep the momentum going, Elliott and Thomas decided to establish a new 'cable scrap analysis' position devoted solely to analysing scrap, tracing its source, and reporting on a daily basis to shift managers and producers. In this way the systematic monitoring and analysis of scrap was institutionalized, and daily, weekly, monthly and annual scrap figures became one of the key internal efficiency indicators of the factory.

The other fast track projects showed similar improvements. For example, the finished goods stock project group had as one of its objectives the control and monitoring of raw materials supply. They decided that if they could achieve a hit rate of nearly 100 per cent, that is suppliers delivering raw materials without fail at the agreed date, they could move to ordering materials 'just in time' and make significant savings in the cost of materials stocks. By the end of April 1992 they had raised the hit rate from just over 60 per cent to 98 per cent, simply by keeping systematic records of suppliers' delivery performance, sending them regular letters telling them of their performance, and advising them that this would be taken into account when the company came to review their contracts. Similar improvements were achieved in finished goods stock, whose value was reduced by 25 per cent between September 1991 and April 1992.

In contrast, the work-in-progress (WIP) project team decided that their main problem was the lack of accurate real time information to measure work-in-progress. The task they set themselves was to programme the plant operations management system (POMS) software to generate this information and to make it easily available on screen for all managers and producers (on POMS, see Chapters 4, 5 and 10). This they achieved. By the spring of 1992, it became possible at the touch of a button to find out exactly when the cable on each bobbin had been produced, what its value was, what percentage had been produced in the previous three days, and thus how quickly production was proceeding through the factory. Setting up such an automated WIP control system was a precondition for moving to a new and more focused form of 'just-in-time' production. This no longer meant, as originally planned (see Chapter 4), scheduling cable production in real time in direct response to specific customer orders. Instead they created a JIT system to meet their operational needs in which each stage of production (metallurgy, extrusion, laying up, packing or rewinding) was closely coordinated and computer monitored. The aim was to use the information to reduce to a minimum the time spent between each stage, and thus the amount of expensive raw materials tied up in WIP.

The WIP project was particularly significant because its team leader – chosen by Elliott after consultation with the chief systems engineer – was Stuart Wood, the systems engineer who had recently been put in overall charge of POMS implementation and development. As project team leader, he played a key role in developing POMS software for real time control of work-in-progress. More importantly, the successful outcome of the project showed the potential of POMS when a sympathetic and knowledgeable in-house systems engineer worked together with experienced and committed production staff on concrete problems. The incremental development of the capability of POMS, which will be discussed in more detail in Chapter 10, came to be the most positive example of 'continuous improvement' in 1991 and 1992.

The first three TQM projects demonstrated many of the fundamental features of good TQM 'fast track' initiatives. They dealt with priority problem areas which were crucial to the achievement of the company's wider business objectives; they had clear goals, procedures and performance measures; and they involved a team approach to problem-solving. Their work was also extremely visible to everyone in the factory, most especially the work of the WIP project team, which converted the POMS data into a neon light strip system at the main exit from the factory. This updated the real time monetary value of WIP every hour. Below the neon light system were charts which plotted the month-by-month progress of the three project teams towards their stated objectives. TQM is often described as 'management by display' and the visibility of the project team objectives helped focus the Aberdare culture on a more precise and measurable set of quality improvement objectives.

## Phase 2 of TQM at Aberdare: quality improvement groups

The three fast track projects at Aberdare had three main drawbacks. First, they were time-limited, with teams formed to achieve particular targets and then dissolving once these had been reached. Second, they only worked on problems identified by management. Finally, they only involved a small number of employees. The next phase in TQM, encouraged by divisional management and implemented from March 1992 at Aberdare, involved the creation of multi-disciplinary quality improvement groups or QIGs. These were defined in the company-wide *TQM Review* as 'a team of employees . . . empowered by divisional management to devote part of their working week to improving their own process'. Two new quality improvement managers were appointed at divisional level to help monitor the TQM programme and help train employees in TQM awareness or problem-solving tasks, as well as in TQM techniques such as PICA boards and statistical process control (see below).

For the first time at Aberdare, the QIGs gave non-management staff the institutional backing and resources to help set their own priorities for the improvement of their own work processes. The first Aberdare QIG, set up on the initiative of shift manager David Norton, drew in all the producers from the packing section on his shift, together with one systems technician and two members of the maintenance department. They decided at the first meeting to try and improve the inconsistent performance of one machine that was causing immense frustration to them as well as inhibiting production efficiency.

Having all attended a divisional TQM awareness course, they decided early on to use the PICA board technique. This was an idea which had been developed in Japan by the Sumitomo Electric Company. The 'process improvement causal analysis' (PICA) board, which is displayed in the machine or process improvement area under investigation, is basically a cause and effect diagram. It records failures in particular processes, the reasons for failures, and what is being done to ensure that they do not recur. PICA boards are normally designed so that there is ample space for anyone to write or pin up their own problems, suggestions or comments. Under the procedures laid down by divisional quality managers, all suggestions for improvement and comments attached to the PICA board had to be dated and labelled by the QIG team as either 'under consideration', 'under trial', or 'now part of standard practice'. In short every suggestion had to be considered seriously and feedback given on the outcome within a specified time period. This approach also allowed other shifts to participate in the work of the QIG.

Between March and October 1992, the packing QIG identified a series of problems with packaging materials. Having first analysed their possible causes, the team contacted one of the materials suppliers to suggest certain modifications. This resulted in small but not insignificant savings on material costs as well as a less frustrating machine process for the workforce. On their own initiative members of the QIG maintained contact with the supplier, visited his factory to discuss other problem areas, and suggested further quality improvements in the material leading to greater machine effectiveness. By December 1992 output from the problem machine had improved by around 4 per cent per shift.

A second QIG project group in packing carried out an assessment of a new cardboard carton design. They sent samples to selected customers, and then used the replies to develop detailed proposals as to how the design might be improved. By the end of 1992, three QIGs were in operation in the packing and rewind section, and others were beginning to spring up in different parts of the site, including in the new PVC plant and in the office area. A third group set about applying SPC techniques to machines in the metallurgy section. Given the complex statistical calculations that needed to be done on this project, it was coordinated by the site's quality assurance manager, a science graduate with a high level of mathematical and computer skills.

By the end of December 1992, TQM – in its many guises – had become an institutionalized yet informal system of team collaboration in problem-solving. It also entailed what one producer called the 'real involvement' of employees in the day-to-day conduct of work. Company and site-wide leadership in setting up the TQM programme had been translated into employees taking responsibility for continuous improvement of machines, products and their own working environment.

## Summary

In this chapter we have explored the question of quality management in the Aberdare plant, concentrating in particular on the launch of the company-wide TQM programme in early 1991 and its implementation at Aberdare in 1991 and 1992. A basic level of quality management is a legal requirement for any company producing electrical products. However, the introduction of a total quality management programme in 1991 was intended to be something much wider, namely the stimulus for addressing a number of long-standing issues such as declining profitability and the need to improve customer satisfaction and employee morale.

In many respects a culture of 'soft' TQM – team working, problem-solving, employee involvement – already existed at Aberdare. However, the company-wide TQM programme gave it a much harder focus, with systematic objectives, clear priorities and explicit procedures. The first stage of implementation involved the establishment of 'fast track' projects in three high priority areas. The handpicked team members – generally a mix of line managers, specialist staff, and non-management staff – took part in regular problem-solving sessions over a period of around six months, and each project achieved major quantifiable improvements.

However, the projects were time-limited, worked exclusively on management-defined problems, and involved only a small number of staff. The second, and in some ways, more important stage of the programme involved virtually all staff on site in more permanent quality improvement groups, working on problems identified largely by staff in their own work areas. By the end of 1992 TQM at Aberdare had become an institutionalized yet informal system of team collaboration and quality improvement.

## Questions for Discussion

1   Analyse the relationship between Pirelli's TQM initiative and wider changes in the company's corporate and business strategy.
2   How far and why did the company adopt or modify the seven main TQM principles outlined in the chapter introduction?
3   Give an evaluation of quality management at Aberdare prior to the introduction of the company-wide TQM programme.
4   Discuss the strengths and weaknesses of the first phase of the TQM programme at Aberdare.
5   Assess the main achievements of the TQM programme at Aberdare by the end of 1992.

# 10

# Leadership in Context

## Thematic Focus: Theories of Leadership and Leadership Behaviour

In Chapter 7 we noted Kanter's distinction between two different kinds of individuals – or groups of individuals – with the organizational power and resources to influence the management of change (see Kanter, 1983: 362). First there are the senior executives, who are able to mobilize support for a strategic decision to innovate as well as set the context – in terms of organizational culture and organization structure – for the implementation of change. Second, there are what Kanter calls the true corporate entrepreneurs, those who champion or drive forward the implementation of change and, by adapting strategy during implementation, shape what it turns out to mean in practice. Underlying both these ideal types is an assumption about the nature of leadership in organizations.

Until the 1940s, research into leadership was dominated by the *trait* approach, the search for identifiable and measurable characteristics of successful individuals (see Bryman, 1992: 2–4; Huczynski and Buchanan, 1991: ch. 19). The failure of this approach to establish a single set of general traits which are effective in all situations led many researchers to dismiss the whole idea. A minority (for example Bryman, 1986) has continued to argue that there are identifiable connections between particular personality traits and leadership success. However, more recent studies have placed a greater emphasis on the importance of the *context* in which leaders operate. Indeed, this is now seen as perhaps the most important variable in explaining why certain types of leadership behaviour are successful and others not.

There are many different elements of context: industry or sector, product market, organization structure and culture, size of company or establishment, political environment, and so on. However, two elements are of particular relevance for our study of the management of technical and human resource change. The first is the character of the staff who work with or for the 'leader', and the extent to which they are prepared to recognize the exercise of leadership as legitimate. Seen in this way, leadership is a property of a relationship, not of an individual. Second, if we assume that context is important, it follows that some types of leadership behaviour and style may be more appropriate at certain stages in processes of change than at others.

How might we define leadership? Much of the debate has taken its cue from the three features identified by American management writer Peter Drucker in the 1950s (1955: 195): *lifting vision, raising performance* and *building personality* beyond existing limitations. John Adair, the prolific UK writer on leadership, has identified three similar features (Adair, 1990): *direction* – close to Drucker's vision and the defining feature of a strategy (see Chapter 1); *team-building capacity* – close to Drucker's personality and performance features and involving an ability to motivate and communicate; and *creativity* – the ability to innovate. Many elements of these two approaches are incorporated in Bryman's concise working definition: 'the creation of vision about a desired future state which seeks to enmesh all members of an organization in its net' (Bryman, 1986: 6). Leadership thus differs from management, with its more technical 'preoccupation with the here-and-now of goal attainment' (1986: b).

One of the most influential conceptualizations of leadership behaviour is that of Fleishman (1953a, 1953b) derived from research with colleagues at Ohio State University in the late 1940s. He distinguished between two different basic types: task-oriented and employee-oriented leadership. Task-oriented or job-centred leadership behaviour gives priority to the setting of objectives, initiating action, making expectations clear, and setting performance targets. In contrast, employee-centred leadership is a more needs- and relationships-oriented behaviour in which the leader (or leadership team) exhibits an ability to listen to staff, support them, motivate them, and enhance their self-esteem. Thus, as with HRM (see Chapter 6), leadership behaviour can be differentiated into hard and soft versions, one concerned with tasks and outcomes, the other with people and processes.

The Ohio studies suggested that the most effective leaders are those who both get the job done and maintain good team relationships, that is, marry task- and employee-orientation. Not surprisingly, this does not always work out in practice. Indeed, while most people prefer considerate and employee-centred rather than task-oriented leaders, a 1987–8 survey of 1000 senior and middle managers in Europe found that the typical company chief executive was a lonely, ambitious, strong-willed autocrat, making lone decisions and motivated by power and money (see Hucyznski and Buchanan, 1991: 492).

Much of the recent debate has focused on different types of leadership (and in some cases management) *style*. The most influential early research was the study by Rensis Likert at the University of Michigan, which identified four main styles of leadership in work organizations (Likert, 1961): exploitative autocratic, benevolent autocratic, participative, and democratic. Carnall (1990) has developed a modified and updated version of the four styles, redefining them as: participating, delegating, selling, and telling. All the evidence suggests that there is no one 'best' type or style of leadership for all situations. The effectiveness of a particular leadership style depends largely on context.

However, if an organization wishes to emerge from the implementation of strategic change with a workforce which is committed to innovation rather than simply compliant with management instructions, then some kind of participatory or delegatory approach is likely to be the most appropriate (Huczynski and Buchanan, 1991: 518). In this context, the distinction between 'transactional' and 'transformational' leadership is fundamental (see Burns, 1978; Bass, 1990; Tichy and Devanna, 1990). Transactional leadership involves an exchange between leaders and subordinates, a kind of contract under which both parties meet their limited obligations but neither goes beyond them. This kind of leadership generates compliance rather than commitment, and does not bind the parties together 'in a mutual and continuing pursuit of a higher purpose' (Burns, 1978: 20). This latter is the key characteristic of transformational leadership, in which leaders aim to engage their staff as whole people, and not simply as individuals 'with a restricted range of basic needs' (Bryman, 1992: 95). Both parties raise each other's aspirations and levels of motivation around a clear sense of higher purpose.

In the following discussion we will use these various distinctions and concepts to assess the leadership of four very different individuals involved in the Aberdare project.

---

## An Assessment of David Regan's Leadership

The context within which David Regan began working for Pirelli General in January 1984 was fraught with risks. His task was to build, equip, staff and implement an experimental factory with a degree of machine innovation and computer automation that existed in no other cable factory in the world. Some of the technical innovations would certainly not have been chosen if the original goal had been to set up the most production-efficient and profitable factory in the short term. They were chosen because sector management in Milan wanted the UK company to conduct an experiment on behalf of the group and to learn from its failures and successes. The company had no 'home-grown' manager whom it felt it could appoint to the post of project manager for the new plant, and so it brought in Regan from outside.

Throughout his seven years in charge Regan personified the Aberdare ideal: the achievement of lowest cost production through innovations in technical and (increasingly) human resource management. The loyalty and commitment he generated among his staff sustained them in the belief that they would achieve the ideal, even in the darkest days of 1988 and 1989 when it seemed as if things would never come right. His greatest talent was an ability to motivate and gain the commitment of the people who worked for him to achieve the ideal. He was a transformational leader, described by various senior and middle managers in the following words: 'inspirational', 'he made me feel I was someone special', 'we were trying to achieve things for him as much as anything', 'we'd never have got where we are today without him', 'John the Baptist', 'he gave me complete authority and unswerving loyalty, I'll never forget that'. Many of the non-management staff felt the same way, although from 1989 onwards he became more remote from the shopfloor, spending much of his time outside the plant with customers or attending corporate meetings in Southampton.

In one particular sense, he was also extremely unlucky. Just at the time in 1990 (albeit 18 months later than initially projected) when the factory began consistently to meet its production output targets, the UK economy – and the construction industry on which so much of the building wires product market is dependent – entered the longest recession since the Second World War. At the time, most economic commentators and all government ministers argued confidently that the recession would be short-lived and that the 'green shoots of recovery' (to quote the then Chancellor of the Exchequer, Norman Lamont) would soon blossom into a period of sustained economic growth. Much of Regan's concentration on production output in 1990 and 1991 can be explained by his belief that the market for building wires would soon be expanding in line with a widely-predicted general upswing in the economy. It is a salutary reminder of the influence of the business cycle, and political predictions about its immediate future, both on commercial strategies and the careers of individuals.

In terms of Peter Drucker's three main features of leadership, David Regan certainly had the ability to 'lift vision' and to 'build personality beyond its normal limitations'. Meeting the site's production target in 1990 was a remarkable team performance of the whole factory given the problems that had been experienced over the previous three years (for many of which, of course, he as overall project manager had to bear responsibility). The qualities and context of his leadership are well captured in the following tribute from one of his senior management colleagues, given in 1991 just after Regan left Aberdare:

> Give the staff here a chance to put the volume through, there's nothing that can touch them . . . Go into the black and everyone would be ten foot tall . . . The level of expectation of applicants who applied for jobs here [in 1987 and 1988] was so low it was abysmal. If you take that into account, you have a quantum leap. It was the magic of Regan. What they've achieved is incredible. They haven't come out of the caves, but into the year 2000 already.

However, Regan's strengths were also his weaknesses. His leadership style was much less suited to the context within which the Aberdare site had to operate in the autumn of 1991. He was pre-eminently a person of vision and inspiration, less one of detail and quantification. The quality of his leadership was more employee-oriented than task-oriented. He established few rules, rarely kept records of meetings, and trusted both management and non-management staff to be self-supervising. But he did not create an environment in which discipline, formality and systematic monitoring could be applied when they were needed.

One example of this regime emerged in November and December 1991 when the new energy cables divisional accounts manager ordered an inventory check of finished goods stock and found a significant discrepancy between actual and 'book' stock. It is arguable that the discrepancy arose from a cumulative failure to monitor stock within the division – going back even before Regan's time in charge – as various distribution depots closed and old stock was moved to new ones. However, the fact that no systematic stocktake was done between 1987 and 1991 was indicative of a general lack of priority given to such matters.

However Regan's overall contribution might be judged, the company decided that it needed to replace him as Aberdare site manager in the autumn of 1991. He had been there five years (seven if the two years' preparatory work in Southampton are included), had given it his all, and was beginning to run out of steam. More importantly, what he had to offer was not really what the factory needed at that stage in its development. It needed a new kind of leadership in a new context, one still committed to the innovative technical and human resource goals of the site, but more prepared to give priority to internal efficiency objectives and the nitty gritty of machine, process and systems improvements. This was the context facing the new Aberdare site manager, Michael Elliott, just one year after he had arrived at Aberdare as operations manager.

Elliott concentrated on three interrelated ways of achieving these goals: TQM, the reorganization of the maintenance department, and incremental development towards a computer-integrated business. We have already looked at Elliott's role in the introduction of TQM at Aberdare. The other two areas provide interesting case studies of the interaction between 'supportive' (see House and Mitchell, 1974) leadership from Elliott and very different kinds of focused, task-oriented leadership – one from a middle manager, the other from a professional engineer.

# Case Study of Leadership in the Reorganization of Maintenance

The one group of operational staff who were not involved in any of the first three fast track TQM projects were the maintainers. This was no coincidence, as the maintenance department had never really established itself as a cohesive and efficient unit. It had suffered from weak departmental leadership and the existence of a small group of staff who were determined to continue with some of the more traditional, low-trust working practices that had been prevalent in the old Aberdare factory.

As we saw in Chapter 7, tensions came to a head in 1990 over two issues: the applicability of the skill module system to the department, and the different treatment of fitters and electricians. These problems were eventually resolved, but still left a fundamental malaise in the department. This only became fully apparent when the original maintenance manager left in the summer of 1991 to be replaced by Martin Parker, maintenance manager at the company's industrial and special cables factories at Bishopstoke.

Except for one year in the late 1970s when he worked as a freelance, Parker had been with the company since the age of 16. Like Elliott he had started as a trainee technician and had worked his way through a range of production and maintenance management posts in various divisions. Between 1988 and 1991 he had gained a reputation as a tough if highly-effective maintenance troubleshooter.

He was approached by Elliott in the summer of 1991 to take over the maintenance manager's job at Aberdare. The two knew each other well, having worked together in the late 1980s when Elliott had been involved in transferring machinery from the Southampton factory to the company's Bishopstoke site 10 km north of Southampton. However, Parker did not accept the post until September 1991. First, he wanted to visit Aberdare to find out for himself exactly what the job entailed. He also still had a job to complete in Bishopstoke, which did not finish until the end of November. For three months in the autumn of 1991, he was effectively working a seven-day week – carrying on his full-time job in Bishopstoke while also visiting Aberdare for between two and three days a week.

Parker was highly impressed by the plant layout at Aberdare and excited by the level of automation, which he regarded as 'world class'. However, he had been warned by Elliott of the problems in the maintenance department, and he soon discovered that there was a complete lack of strategy (direction) about its role in such a high technology factory. He also found an absence of basic management systems in maintenance – there was no budgetary control, no organized system of workload management or stock control – and no maintenance involvement in TQM. Finally, he was rather taken aback by what he felt were the high numbers of maintenance staff (seven non-management staff plus a shift automation engineer on each shift) in what was supposed to be a highly-automated factory.

## A new strategy for maintenance

Within a year Parker had literally changed the whole way the department operated. His first priority was to establish some basic management systems. He began by introducing a budget for maintenance stores and parts. This was controlled and monitored on a weekly basis. Parker instituted regular meetings of the three senior shift maintenance personnel (the automation engineers) to ensure that the department was being managed consistently across all three shifts. In addition he charged each of them with additional responsibilities, setting tasks and clear deadlines by which he expected them to be completed. He also established a new workload management system.

He then set about designing and implementing a whole new strategy for maintenance. This had three main objectives: to transfer routine and minor maintenance activity to production staff; to establish the maintenance department as a small, highly-trained unit of multiskilled specialists; and to shift the focus from time-based planned maintenance to a greater emphasis on 'predictive maintenance'. By July 1992, he had seen the size of the maintenance department reduced from 25 to 16 in the wake of the redundancy programme, and had helped establish three new posts of producer/maintainer, which became the prototypes for a longer-term programme to upgrade the maintenance skills of producers (see Chapter 8 for details). His most fundamental innovation, though, was the introduction of a whole new philosophy for the department centred around the idea of 'predictive' maintenance.

When he arrived at Aberdare, much of the maintenance effort was still 'corrective', that is, correcting faults after they had occurred on a fire-fighting basis. At that time, 'preventive' maintenance consisted of a mixture of routine operator maintenance (such as cleaning and lubricating mechanical parts on machines) and planned maintenance in which main-tainers, assisted by producers, would take machines out of production and give them a thorough check and overhaul (akin to a routine car service). Parker found that the planned maintenance programme, which had been devised by manufacturing manager David Thomas, was basically regulated by the calendar: every 3, 6, 12 or 24 months particular machines would come up for a service. But no one appeared to be comparing the fault records of different machines, examining the down times, or even checking to see how many hours they had been running between services.

The first prerequisite for dealing more efficiently with planned mainten-ance was to set up an information collection and processing system which could generate data on these matters. This task was taken on by the systems engineer in charge of the plant operations management system, Stuart Wood (see next section for a discussion of Wood's wider role). By the summer of 1992, POMS was able to generate works orders (maintenance tasks to be done) automatically, provide historical and real-time data on downtime and total running time for each machine, and thus identify which machines would need to be taken out of operation for a full maintenance check.

## Predictive maintenance

A few months before coming to Aberdare, Parker had read in the trade press about a new maintenance technique called predictive maintenance. He was already in the process of introducing it at Bishopstoke when he was transferred to Aberdare. Predictive maintenance is a form of vibration analysis or machine condition monitoring. A number of test points are identified for each machine (usually with the help of equipment suppliers), vibration tolerances are set for each test point, and alarms identified when vibration levels go beyond a certain point. The test points are then checked on a regular basis using a data collector (a black box mini computer which takes readings), an accelerometer attached to the data collector which is applied to the test point like a stethoscope, and a computer plus software into which the data from the data collector are downloaded for analysis. The computer then produces a graph – like that of a heart beat on a cardiograph readout – which plots the vibrations and movements above and below the set tolerances.

Soon after he started coming to Aberdare, Parker sent the acting maintenance manager, Peter Mandell, on two courses in predictive maintenance. Over the following six months all maintenance staff either attended a basic course or were given training by Mandell, who was given overall responsibility for setting up the test points, carrying out the tests and analysing the results. The new approach soon paid off spectacularly. The analysis helped identify a major hidden fault in part of the new PVC compound plant which would have caused untold damage if it had 'blown' and may well not have been discovered until the costly offending part was out of manufacturer's warranty. By August 1992 Mandell had established 1200 tests points in the factory, and by December 1992 2000. By the beginning of 1993 the number of corrective faults had reduced significantly.

### An assessment of Martin Parker's leadership

Martin Parker left Aberdare in October 1992, returning to South Hampshire to take on another troubleshooting job in the company. He had been at Aberdare for less than a year and achieved an enormous amount in that short time. He had shown direction and creativity (two of Adair's three elements of leadership), and lifted vision and raised performance (two of Drucker's three elements). He had also been a highly efficient manager and administrator. But he also left behind a maintenance workforce with low levels of morale.

Parker's approach to leadership had been highly directive and task-oriented. It was focused on outcomes and the application of highly rationalistic and precise methodologies. He was rarely seen listening or talking to staff, who felt he had done little to raise their motivation and self-esteem. He was highly self-motivated, efficient, well-organized and demanding, keeping everyone on their toes. In many ways, the 'context' of the maintenance department demanded this and Parker provided what was needed. He left the maintenance department a much more focused and efficient place.

# Case Study of Leadership in Promoting Computer Integration

### POMS leadership: the role of Stuart Wood

A very different kind of task-oriented leadership was provided by systems engineer Stuart Wood between 1990 and 1993. He was not a leader of the 'grand gesture', more a 'quiet innovator' (Kanter, 1983: 354). Wood was a mechanical engineering graduate with a strong interest in computers who had joined Pirelli General straight from university in the summer of 1985. Towards the end of his first year as a graduate trainee at PG, he was seconded to the Aberdare 'systems' team under David Wolstenholme (see Chapters 4 and 5) and given responsibility for producing a report on 'computer disaster recovery', in other words, what would happen if the Aberdare CIM system crashed, how could it be retrieved, how much would it cost, and so on. Fascinated by the experimental use of computers in manufacturing promised for the new factory, he kept pressing his superiors to be taken on permanently at Aberdare. Finally, in April 1987, he was appointed to the systems engineering department, moving down to Aberdare in November 1987.

For the first three years, he was very much in the computer backroom of the department, developing accounts codes for the different products and helping to set up much of the manufacturing management system data base. He also worked on the data base for POMS, which involved giving codes to every machine and every bobbin input and output stand, all the bobbin store positions, and so on. At this time David Wolstenholme called him, half-seriously, the 'site data manager'.

In October 1987 he visited the headquarters of the POMS software supplier, SEIAF, to attend the 'final test' of the full POMS system. On his return he helped build up manufacturing management system files which would integrate automatically incoming customer orders, the design specifications of particular products, and bills of materials. His colleague Hugh Marr was doing similar work on the POMS data base. Throughout this phase, there was little or no regular contact between production managers and systems engineers. It was very much a matter of parallel rather than collaborative working – a classic 'polarization of expertise' (see Chapter 4).

## The changing context for the development of POMS

A number of events caused this to change and opened the way for a flowering of POMS development initiatives from late 1990 right through until 1993. The two most important were the increase in machine reliability – a precondition for POMS to operate successfully – and the decision in early 1990 to end SEIAF development work by the end of March 1991 (see Chapter 5). Then, in February 1990, David Wolstenholme died tragically. With Gordon Peters still 'on loan' from the corporate technical department, it was crucial for the Aberdare management team to identify their own staff to take over full-time responsibility for POMS from SEIAF. Wood's colleague Hugh Marr had had three hard years on POMS, but had become immensely frustrated with his lack of authority to make changes to the software while SEIAF was still formally in charge. By 1990 he was only too willing to accede to the chief systems engineer's suggestion that it would be better for someone new but knowledgeable to take over POMS when SEIAF's contract ended. Wood was assigned to the task.

In the summer of 1990, Wood was sent to Genoa for two months to learn in more detail about the SEIAF system and POMS programming. By the time Michael Elliott took over as Aberdare operations manager in September 1990, Wood was both knowledgeable and full of enthusiasm for what could be done with POMS. But he found himself operating within a manufacturing subculture in which production managers (and their staff) were more interested in achieving production output targets – if necessary by using existing manual and paper-based operating systems – than in new POMS developments or POMS trials (see Chapter 5). On one famous occasion in mid-1990 there was a bomb scare in the factory during the night, and evacuation took an extraordinarily long time to complete. The next day, the word went round that it would have been much easier to get a quick evacuation if it had been announced as a 'POMS scare' rather than a bomb scare!

### Persuading managers to use the existing POMS data

At this moment in time, the fact that Elliott was new to Aberdare was a great advantage. His predecessor as operations manager had been a staunch advocate of POMS when many around him – and many at very senior levels at corporate HQ – would have preferred to see it dropped altogether. But Elliott arrived in the autumn of 1990 untainted by any previous experience and could ask 'naive' questions about why things were as they were. Knowing Elliott's openness and willingness to listen, Gordon Peters persuaded Wood to see him to explain in detail what could be achieved with POMS and what were the main barriers to change.

Elliott, who had little knowledge or experience of the use of computers in manufacturing, was impressed by what Wood told him, noted the various possibilities for POMS development, and decided to discuss them in detail with manufacturing manager David Thomas, the Aberdare planning coordinator Graham Hall, and the three shift managers. One by one, Elliott went through their concerns, discussing the advantages and disadvantages of various options and trying to gain consensus for some immediate and achievable developments which would be of direct value to the production effort. In this way, he also aimed to win them over to accept new applications of POMS in future.

His first success was to persuade them into using the information which was already accessible through POMS. For example, the existing software already provided an inventory management system, with details of where bobbins were located, what material was on them, and what types and quantities of finished goods stock were available. But the production planning department – which was also responsible for stock control – was not using the computerized data, preferring still to carry out physical checks of work-in-progress and finished goods stock. This took between two and three hours per day and the data were already out-of-date by the time they were collected and collated. Soon after Elliott's arrival the production planners stopped doing physical checks and began to use POMS data instead.

### Redesigning the operator interface

Elliott also encouraged Wood to expand his work on redesigning the operator interface – the display of POMS data on computer screens located next to each machine – to make it more user-friendly. Wood worked on this in 1990 and the first half of 1991. At each stage he discussed the redesign with the users, getting them to look at his computer simulations to see whether the design was appropriate and how it could be improved. An important role in this exercise was played by the three POMS 'systems' technicians. They had been selected in 1990 from among producers to receive additional POMS training and to act as a link between the shopfloor and the systems department. The final impetus for the implementation of Wood's new operator interface system was David Regan's decision, in May 1991, to abolish the use of paper on the shopfloor (see Chapter 9).

At the same time as Wood was putting the final touches to the operator interface, the new TQM programme was being launched across the company and the first fast track projects were being identified. As we saw in Chapter 9, Wood was made team leader of the work-in-progress fast track project, working directly with shift managers, production planners and producers. The momentum for the incremental development of POMS towards its intended place at the heart of a computer-integrated manufacturing system was now becoming irresistible. Success bred success. By the end of 1991 the redesigned plant operations management system was in operation and running to the satisfaction of all concerned.

### The new POMS system in operation

To appreciate what had been achieved it is best to compare the *new POMS* system with the original fully automated design of 1986/7 (*full POMS*, as outlined in Chapter 4) and the interim system (*virtual-POMS*, see Chapter 5) introduced in 1989/90 following the decision to retreat from full automation.

Full POMS was intended to manage all shopfloor operations automatically. There was to be no operator discretion, no choice, just automatically downloaded instructions which the operator would be required to carry out to the letter. With such a system, it was not surprising that no user interfaces were developed, as the communication was intended to be all one way. In addition, little provision was made in full POMS for the automatic generation of management information about how production was proceeding, the cost of work-in-progress, and so on.

1989 was the year of the retreat from full automation. Regan and his senior colleagues had recognized that to achieve a fully automated CIM factory would have required an extraordinarily complex and wide range of alternatives to be built into the software to deal with all possible contingencies, particularly for when things went wrong. By neglecting to build human intervention – whether of operators or managers – into the original specification and design of the system, the company was also wasting a major source of knowledge and experience to deal with unexpected events. Day-to-day manufacturing practice often requires rules to be changed or modified at short notice to meet changing circumstances. Skilled and knowledgeable human beings are uniquely qualified to make on-the-spot judgements about such changes. As Wood expressed it later: 'if human beings aren't allowed to make decisions, there is a lot of processing power out there going lost'.

Under Wood's new POMS design, the POMS 'shopfloor supervisor' (SFS, see Chapter 4) still stores all the planning data in exactly the same way as originally intended. However, it is now the planning staff who decide which orders and jobs they wish to run, not the POMS software. The POMS 'machine supervisor' receives this planning information automatically from the SFS and downloads it, together with the basic information needed to complete the job, onto a POMS terminal next to each machine. The daily plan of jobs to be run on each machine is no longer a rigid set of instructions which drives the production process automatically. Instead operators (and managers) can treat the plan as a 'wish list' which they can modify if they decide that a different order would be more efficient or appropriate under the particular circumstances prevailing at the time. Where previously they had ended up fighting with POMS when they wanted to make modifications, they can now work with it.

Despite this greater degree of flexibility in choosing which items to run, once an item is chosen, new POMS does still establish a clear set of parameters from which it is not possible to deviate. It also simplifies and restricts the interaction between the operator and POMS. The contrast with v-POMS illustrates most vividly the character of the new operator interface. Under v-POMS, operators were able to generate a particular piece of work or call up a particular bobbin by typing in a code. They then had to wait for POMS to respond, which sometimes took a few minutes – occasionally hours. This caused immense frustration all round. Under Wood's new system, all the jobs, bobbins and code numbers are available automatically on screen. Operators now select the jobs and raise the AGV missions for the bobbins from a limited range of specified options by moving the cursor on screen to the option they want and pressing the input button. The operator then requests the set-up information for the job and these appear automatically on screen. Throughout, the POMS machine supervisor eavesdrops on the machine plc and monitors how production is progressing.

Under v-POMS, when operators had completed a particular job, they were asked to type in the code of the bobbin they had been using together with a range of other data about the particular job. Now, virtually all this information is recorded automatically. Operators are also informed automatically which quality assurance tests need to be carried out on each cable product. They are then asked to confirm individually that these have been done and record any defects (or other comments). Only when this has been done are they able to raise an AGV mission to send the bobbin to the next stage of production.

New POMS is more responsive to operator intervention than either full POMS or v-POMS. It is also more secure and resilient. It leads operators through the various stages of production (choice of job, raising AGV missions, setting up the machine, monitoring progress, recording machine faults, carrying out cable quality tests) by a series of forced choice questions or alternatives. This reduces the possibility of human error experienced under v-POMS, where operators had to type in long code numbers. As the actions to be taken are now comparatively simple, new POMS also reduces the amount of time operators need to spend at the POMS terminal or to wait for the machine to respond to their instructions. Nevertheless, the reduced time spent at the POMS terminal is still crucial to the whole operation of the plant.

The new POMS system also enhances the availability, speed and reliability of management information about production operations. In traditional cable factories, the only way of finding out about work-in-progress is by time-consuming physical checks. Even when these checks are done, it is not normally possible to identify how long a particular bobbin has been where it is, nor the monetary value of the work-in-progress contained on it. Now, as a result of the work-in-progress fast track project under Wood's team leadership (see Chapter 9), managers and operators are able to find out at the press of a button the exact location of each bobbin, the exact cost of work-in-progress on them, and how long they have been on the factory floor (crucial for controlling costs and a prerequisite for just-in-time production). Graphs or tables of total work-in-progress can also be generated at the press of a button. In addition, new POMS can supply information about the exact amount of tonnage as well as the cable faults and scrap levels produced by each shift. It also records automatically the number of hours each machine has been in operation, using this information to generate reminders about the need for planned maintenance.

'What has happened with POMS is all down to one man, Stuart Wood.' This was a view expressed by many different members of the Aberdare staff – managers, systems engineers, production operators – when interviewed in August and September 1992. Undoubtedly, his engineering background, his fascination with computer systems, his experience with POMS since 1986, his frustration that the potential of POMS had not been realized, and his commitment to put in long hours to redesign it in consultation with production staff, were all crucial factors in achieving the advances that had been made. But the context and general environment within which he could exercise this kind of technical leadership were also vital.

The contextual factors included: the major improvement in machine reliability from 1990 onwards; Wood's selection in 1990 by senior management as the new systems engineer responsible for POMS; the ending of

SEIAF's involvement which allowed him to make the system modifications he wanted; the advent of Michael Elliott as operations and site manager, who was prepared to look at POMS afresh; the support and advice of Gordon Peters; the introduction of the company's TQM programme; his selection as team leader of the work-in-progress project; the holding of the Pirelli world general wiring conference in Aberdare in June 1991, which motivated David Regan to do away with paper transactions on the shopfloor; and the commitment of POMS system technicians to making it work at a time when the vast majority of producers and their managers would have preferred it to be scrapped. Wood was part of a team, but he was the right person in the right place at the right time – a slightly unusual backroom leader who transformed, incrementally and quietly, the way the factory was operated and managed.

By mid-1992, Wood had also converted the existing WIP and finished goods data into a form that could be used by management accountants, commercial managers and sales staff to calculate the exact costs of raw materials, WIP, and the production of a finished cable in the factory. By the end of 1992, he was busy devising various ways of extending the use of POMS into the management and control of warehouse and distribution operations at the far end of the factory (see Figure 4.1).

By the end of 1992 the computer systems operating at Aberdare centred around POMS were both an aid to manufacture and a comprehensive management information system. But they were more than this. POMS was now the medium through which production planners, production managers and production operators actually managed the daily and weekly workflow through the factory. It was now taken for granted – it had become an indispensable 'part of the wallpaper'. By establishing parameters, standards and quality checks, it was programmed against operator error and, as Gordon Peters put it, 'immune against bad management'. It was the main cohesive element uniting all of the factory's production operations. It was also becoming more central to its commercial, distribution and financial activities. In this sense it could be described as a computer-integrated system of manufacturing management which was on the way to becoming a computer-integrated system of business management. This had been the ideal of Stokes, Regan, Wolstenholme and Peters in the mid-1980s. Wood and Elliott had provided the kinds of supportive, task- and employee-oriented leadership that rekindled this ideal in 1991 and 1992 after the retreat from full automation in mid-1989.

## An Assessment of Michael Elliott's Leadership

In the middle of the redundancy exercise in June 1992 (see Chapter 8), Michael Elliott had announced that he was leaving Aberdare to become site manager of the industrial cables factory in Bishopstoke. When he had taken over as operations manager at Aberdare in September 1990, the plant had been about to achieve its output target for the first time ever. The workforce was already highly flexible, cooperative and committed. But there were still major problems of machine reliability; daily production planning was still paper-based; levels of scrap, work-in-progress, and finished goods stock were very high; the maintenance department was in a poor shape; and POMS was widely regarded as an expensive white elephant which should be scrapped.

By the time he left in September 1992, machine reliability had improved beyond all recognition; POMS had been adopted enthusiastically as the central mechanism for managing production and, increasingly, production planning and logistics; levels of scrap, WIP, and finished goods stock had been significantly reduced; a new PVC plant and warehouse were up and running, and increasingly integrated with production operations; four of the five regional depots had been closed saving substantial overhead costs; a new strategy for maintenance had been devised and implemented; the workforce was leaner and more experienced, if more fearful and with lower morale; and the factory had survived, even though the revised outturn in 1990 and 1991 meant that it was still some way from making an operating profit.

Many of these developments were already in train when Elliott arrived. The leadership of individuals such as Martin Parker and Stuart Wood had been decisive in achieving particular targets in maintenance and computerized management systems. The company's TQM programme had galvanized all grades of staff to develop and introduce a series of machine and process improvements. And yet, Elliott had been the catalyst for many of these changes. He had attempted from the start to find the right balance between continued commitment to the positive elements of the Aberdare philosophy and a more systematic focus on targeted improvements.

Above all, he was effective because he worked continually with and through his middle managers: listening to them, bringing them together, encouraging them, occasionally cajoling them into doing things they did not want to, but also harnessing their enthusiasm to specific ends and then delegating the authority to achieve them. He was also not afraid of taking tough and necessary decisions, for instance in relation to the maintenance department. David Regan was described by his managers as inspirational and a 'managing director' type of leader. Elliott was described as genuine, respected, helpful, approachable and appreciative – more a supportive team builder and team leader. And yet he did give direction to the site, particularly in terms of machine improvements, TQM, maintenance, and POMS. His type of leadership was a mixture of task- and employee-orientation. His strength was to raise the performance of his managers and give direction to their efforts.

In his valedictory message to staff at Aberdare in the June 1992 team briefing, he summarized his view of what he felt the factory had achieved and where its task lay in future:

> The last two years have further demonstrated the commitment and capability that exists at Aberdare and many lessons have been learnt that will be of benefit elsewhere in the cable sector. The Aberdare operation remains streets ahead of the

Pirelli operations in many areas. Unfortunately we always have to live within the constraints of the commercial activity/market-place and must continue to mini- mize costs wherever possible. We need to do this whether we are making a profit or not. Although this is both difficult and frustrating, the Pirelli objective (to achieve sales at the right price with production at the right costs to make Aberdare profitable) remains . . .

On the first of September Gino Brandini takes over as site manager, Aber- dare. . . . Gino is well qualified to progress the two most significant areas within the site manager's control that will assist in moving Aberdare into the future. These are the reduction in material overuse and cost, and the development of different/new materials. Please give him every assistance in achieving the results that we all know Aberdare is capable of.

## Profile of the New Aberdare Site Manager, Gino Brandini

Brandini, an Italian electronics engineer in his early 40s, was no stranger to general wiring or to the Aberdare project. Although his early career at Pirelli (1974–81) had been in telecommunication cable research and development in Italy and Argentina, he had worked from 1981 to 1984 as a production and technical manager at the company's general wiring factory near Bari in southern Italy. In September 1984 he had been summoned back to Milan to work in the sector technical department to help coordinate the upgrading of Pirelli's building wire operations world-wide.

His first major assignment was to assist David Regan with the Aberdare project, becoming part of the small project team (with Regan and Peters) which did much of the early work on machine systems for the new factory. In fact, he spent nearly 60 per cent of his time in 1985 and early 1986 working on the project, travelling to Southampton almost on a weekly basis. Soon after the Aberdare project work began, he became involved – although to a lesser extent – in the equivalent building wires project team in Spain, led at the time by Felipe Martinez. Then in 1986 he was given direct responsibility for improving and upgrading the company's Italian general wiring factory near Bari. In 1988 he joined a similar project team attached to the group's factory in Canada, and in 1989 he became involved with a similar team in Australia. In short he became the sector's peripatetic expert in factory improvements and enhancements, specializing in general wires.

By early 1990, he had been very much expecting and wanting to move to a more settled job and, in the autumn of 1990, he was offered – and accepted – the post of corporate technical manager of Pirelli General based in Southampton. This included responsibility for quality and machine im- provements at Aberdare. Then, after less than two years, he was offered the job as Aberdare site manager. He accepted with alacrity. He knew the factory well, had spoken at length with Michael Elliott, and they were united in their understanding of the main priorities for the next 18 months to two years.

The first priority, already mentioned in Elliott's valedictory team brief, was to reduce material usage – something Elliott had not really begun to tackle systematically until late on during his time at Aberdare. Second, the site needed to develop new markets for existing PVC compound. Brandini saw himself as exerting very much a task-oriented and technical leadership, concentrating on these two areas. Socially a shy man and concerned at possible language or communication barriers, he was quite happy to let manufacturing manager David Thomas and personnel manager Michael Hawke take the lead on issues of employee relations and human resource management.

Brandini was also aware of another important external role he could play at this stage in the development of the factory. With his intimate knowledge of Pirelli's UK, Italian, Spanish, and more general cable sector operations, he intended to be a major participant in discussions about closer collaboration between the company's various European operations. Above all he hoped to raise the status of Aberdare in the sector and group, ensuring that it would be recognized as a highly-efficient, top class factory which would play a central part in future developments within Pirelli at a European level.

Here, his Italian background and connections, and intimate knowledge of many of the group's general wiring factories world-wide, were likely to be major assets. With the advent of the European Single Market on 1 January 1993, and increasing international competition from low-cost countries, it was unlikely that cable production and markets would remain the largely national activities that they had been in the past. In this context, Brandini seemed to have the right qualifications and experience to lead a new phase in the factory's short history.

## Summary

In this chapter, four different examples of leadership behaviour in four different contexts were examined. First, David Regan's leadership between 1984 and 1991 was characterized as a visionary and employee-centred one which inspired and motivated staff. It also helped create an organization culture conducive to team work, commitment and involvement. However, in the changed context of the early 1990s, he was replaced by Michael Elliott, who personified a continuing commitment to the technical and human resource goals of the site, while giving greater emphasis to the achievement of more quantifiable production efficiency objectives. The implementation of the TQM programme at Aberdare exemplified Elliott's new kind of task-oriented and participative leadership. He also encouraged two members of management and professional staff to give task-oriented leadership in two specific problem areas: maintenance and computer integration.

During a period of only 12 months maintenance manager Martin Parker transformed the whole way his department operated, establishing basic systems of management control and also giving maintenance at Aberdare a completely different sense of direction. His mixture of 'delegating' and 'telling' was highly effective in achieving new methods of working and quantifiable outputs, although it was generally less successful in generating employee motivation and morale. In contrast, systems engineer Stuart Wood was a quiet innovator – working with production, planning, maintenance, accounts and warehouse staff at different levels to realize a major incremental enhancement in computer integration of management systems across the site. By the end of 1992, POMS had been transformed in the eyes of staff from an expensive hindrance to an indispensable mechanism of day-to-day management. It was well on the way to becoming what had originally been intended, a computer-integrated system of business management.

---

## Questions for Discussion

1 Compare and contrast the strengths and weaknesses of David Regan's and Michael Elliott's leadership of the Aberdare project.

2 To what extent do the 'trait' and 'context' approaches help us understand and explain the different effects of Regan's and Elliott's leadership behaviour?

3 Assess the different kinds of task-oriented leadership provided by Martin Parker and Stuart Wood.

4 To what extent does the Aberdare story confirm the importance of the distinction between leadership and management? Give examples to illustrate your answer.

# 11

# A Happy Ending?

Staff at the Aberdare site spent much of 1993 extending the range of cost cutting and performance improvement measures already started in 1992. By December scrap levels were down on the previous year by 20 per cent, work-in-progress by 25 per cent, and finished goods stocks by 33 per cent. In 1991 the factory had been bottom of the European league of Pirelli general wiring companies on three out of the four key production efficiency measures: raw material stocks, work-in-progress, finished goods stocks, and scrap. By the end of 1993 it was in the top two on three out of the four counts.

Improvements in efficiency and plant utilization were also made in other areas. For example, the quality improvement group in metallurgy (see Chapter 9) used the data from statistical process control investigations to save substantial sums on copper costs. The PVC compound plant was able to increase capacity by 40 per cent. Cuts in staffing levels following the 1992 redundancy exercise (see Table 11.1) led to significant full-year savings in 1993. This was on top of savings from the consolidation at Aberdare of four of the five regional distribution depots. Commercial operations within the Energy Cables Division had become even more focused, so that there was now one specialized sales group dealing exclusively with customers for general wiring and smaller industrial cables. A new commercial strategy had been developed which extended both the product range and customer base of the factory.

The overall outcome of these changes was awaited with trepidation by the Aberdare workforce. In February 1994 it was announced that, in 1993, the Aberdare site had made a 'positive contribution' to the division for the first

**Table 11.1** *Core production and professional engineering staffing levels at Aberdare, 1987–93 (31 December)*

|      | Producers | Maintainers | Systems engineers |
|------|-----------|-------------|-------------------|
| 1987 | 19        | 8           | 7                 |
| 1988 | 39        | 11          | 8                 |
| 1989 | 79        | 25          | 13                |
| 1990 | 87        | 19          | 8                 |
| 1991 | 82        | 19          | 7                 |
| 1992 | 60        | 11          | 6                 |
| 1993 | 59        | 11          | 4                 |

time, significantly exceeding its Manplan target. It was cause for celebration. One of the senior managers said to me enthusiastically: 'I think your book will have a happy ending'.

However, by the end of 1993 the Aberdare site was still some way from being a commercial success. Only when the site is able to increase its positive contribution to a point where it covers its central service costs – to the division, the national operating company, and the multinational group – will it begin to make a 'bottom line' profit as a business in its own right. It is a measure of its improvement and improving confidence that, by the summer of 1995, there were few people at Aberdare or in the wider company who spoke any longer of 'if', but only of 'when'.

Rather embarrassingly for Brandini, many of his European colleagues attributed the improved position and outlook at Aberdare to his own influence on the plant since he had become site manager in September 1992. However, most of what he had been able to achieve was a direct result of the foundations laid by David Regan, Martin Dawson, Michael Elliott and the rest of the staff at Aberdare, together with the success of the new commercial strategy put in place under the guidance of the new managing director, Graham Phillips, who had replaced Felipe Martinez in January 1993. Brandini's own contribution had been to tighten up on material usage and general housekeeping, help secure additional customers for PVC products, and be more active in selling the success of the site – concentrating on its 'hard' production efficiency achievements since 1991. At the same time he never lost an opportunity to point out to his European colleagues the importance of the 'softer' achievements at Aberdare – the high levels of flexibility and skill of the workforce and their practical (and profitable) commitment to continuous improvement of all aspects of site operations.

In April 1994 a symbolic seal was set on the new esteem in which the Aberdare site was now held within the wider company. The Pirelli group house magazine, *Fatti e Notizie*, sent a member of its public relations department to Aberdare to prepare a feature article on the site. It was the first cable factory outside Italy to be profiled in the magazine. In the article POMS and the AGVs featured prominently; savings in copper usage and scrap were duly noted; surprise was expressed that, in a cable factory around 100 m long, it was possible to see a number of machines working in perfect solitude without any staff in attendance. But for the visitor from Milan, the 'most impressive automation area' was the new warehouse:

> Here the technology is one of the future. The finished product, assembled and packed, is delivered to a big machine that automatically collates it and stores it in the high bay area. Now a pallet is coming in: the crane takes it and very swiftly gets it up to a height of fifteen metres and deposits the product in a pre-arranged cell. Some minutes later the crane very swiftly retrieves a pallet and takes it out for distribution. (Pirelli, 1994)

The article noted that this high level of automation – together with the outstanding performance of the site's new haulage contractor – ensured that the site now had a guaranteed delivery time of 24 hours for the whole of

Britain. Customer satisfaction was also reported to be at an all-time high. Six customers were telephoned every working day to check whether they were satisfied with the service they were getting. In the first three months of 1994 the satisfaction rate was consistently around 99 per cent.

In 1988 the national UK press had been greatly impressed by the technical and human resource innovations they had perceived at the Aberdare factory. As we have seen, these were a promise rather than a reality. By 1994 not everything had been achieved as intended in the original blueprint. There had been delays in project implementation. Actual costs had far exceeded the estimate put to the board in June 1985. However, taking into account inflation, the addition of the new warehouse, and a number of other incremental innovations, the total costs were not so radically different from the original estimate of around £20 million made by Paul Bamford in early 1985.

The extreme forms of fully automated, just-in-time production had been jettisoned. There was now a balance, not only between automation and human intervention in the production process, but also between reducing finished goods stocks to the minimum and the need to meet all customer orders within 24 hours. The plant operations management system, which had been scaled back and almost marginalized in 1989, was now the heart of a computer-integrated system of information and business management across the site. Machine efficiency and reliability were high and still improving; the flexibility and skill of the workforce likewise. Levels of work commitment were still high, although morale was still affected by the after-effects of the redundancy exercise and the years of failing to make a profit. Customer satisfaction with service was around 99 per cent. A new commercial strategy was now in place which gave the company a stronger and broader position in the market-place.

Over a period of 10 years, from 1984 to 1994, the Pirelli group had shown remarkable – some would say excessive – loyalty to a brave experiment. Some six years after the plant had been formally opened, it looked like this loyalty was about to pay dividends. This was not before time. After what the *Financial Times* called 'the nadir in the company's 122-year history' in 1992, a new senior management team had taken over at cable sector and group level in Milan. Between 1992 and mid-1994, the group's debt had been reduced by half, the number of factories from 102 to 80, the workforce from 50,000 to 40,000, and more than 600 managers had changed positions (all figures from the *Financial Times*, 10 August 1994). Against this background it is improbable that the group would have been prepared to tolerate the continuation of Aberdare as a loss-making operation.

In March 1994, a number of trees were planted at Aberdare. They were offcuts from trees planted near the old Southampton factory. They had been donated by Pirelli Italy in 1914 when the factory had first opened. For the writer of the article in *Fatti e Notizie* (Pirelli, 1994): 'they represent the continuation of the past into the future'. The management and workforce at Aberdare would dearly like to believe it.

# 12

# Conclusion: Learning from the Management of Strategic Innovation

The main part of this chapter will review the themes of the book and relate them to some of the main threads of the Pirelli Aberdare story. The chapter concludes with some general reflections on managing strategic innovation in the 1990s.

## Thematic Conclusions

### The idea of strategy

As noted in Chapter 1, it is now commonplace to criticize a 'decisionist' model of strategy which regards strategic innovation as the outcome of single-event decisions by top managers, a linear process in which strategy formulation is followed sequentially by strategy implementation. More recent studies have suggested that strategy is a multi-layered phenomenon, an emergent property embedded in organization culture and evolving incrementally out of the ideas and actions of people at many different points in the organization structure. In multinational organizations, in particular, the achievement of consistency between different levels of strategy and structure – corporate, business, and operational – is highly complex, a consistency which needs to be balanced by the creation of space and encouragement for innovation and change.

In looking at the development of the Pirelli Aberdare project, single-event decisions by top managers played an important role in strategy development, for example the far-reaching decision of the Pirelli General board to approve the closure of the Southampton factory and to invest in the construction of a new experimental factory. However, this particular decision emerged out of many separate initiatives at different levels of the multinational group, which were in turn fuelled by a series of crises and galvanizing events inside and outside the company. All these events and initiatives culminated in an informal meeting in Southampton from which emerged a strategic framework that established the future direction of the project. This strategy was subsequently ratified by the board of directors.

Some elements of the strategy put to the board – the reduced level of funding for computerized management systems, the omission of proposals for new packaging equipment – were *ad hoc* modifications of what the group managers really wanted and were introduced in order to reduce the overall cost of the proposal and secure the support of the main formal decision-making body, the board of directors. But what emerged from all these decisions, initiatives and modifications was that essential feature of all strategies: a sense of direction.

Without some kind of strategy, the 'deliberate and conscious articulation of a direction' (Kanter, 1983: 294), organizations, departments, teams and individuals are likely to meet every change in their internal or external environments in an *ad hoc*, pragmatic and short-term way. On the other hand, organizations with strategies must also be able to adapt. Otherwise they will be incapable of learning from experience or responding to new challenges and opportunities. The management of change requires both a consciously articulated sense of direction – a strategy – and an organizational preparedness to modify it in the light of experience and changing circumstances.

The most obvious examples of strategy modification at Pirelli Aberdare were the 'decisions' (a) to retreat from full automation and (b) to abandon the idea of full workforce flexibility. These modifications to strategy were born of experience, emerging out of many months of debate between operational managers and professional specialists. These are the people who, in Kanter's words, 'translate strategy – set at the top – into actual practice, and by doing so, shape what strategy turns out to mean' (1983: 354).

## Making and justifying investment decisions

In Chapter 2 two competing theories of the firm, neoclassical and behavioural, were contrasted. These theories of the firm were supplemented by a discussion of three different models of organizational decision-making: the bureaucratic, the professional/collegial, and the political. Behavioural theory and the political model were viewed as particularly powerful in helping to understand large-scale innovatory decisions, where high levels of uncertainty and risk generate greater scope for interpretation, judgement and 'mobilization of bias'.

Chapter 2 followed the elaboration of a series of options for change which were eventually put before the Pirelli General board of directors in June and September 1985. We saw how a group of senior sector and UK managers won support for one particular option – to build an experimental high-technology factory – in which most of the costs would be incurred in the short term and most of the potential rewards gained in the medium to long term. The main alternative option – to refurbish the existing Southampton factory – was rejected, even though in the short term it would have been less risky, less costly, and certainly more profitable. By going back in time over a year prior to the actual decision event, it was possible to trace how one particular manager – with formal positional power in the organization hierarchy and a particular vision of the future – was able to mobilize support in a way that effectively prestructured the board decision in June and September 1985.

The justification of the various options depended initially on a comparison of the relative rates of return on capital employed, derived from a highly technical discounted cash flow analysis. In practice, these quantitative accounting data were but one reference point in a wider judgement. The eventual decision was dominated by a strong belief, almost a faith, in the potential use of computers in manufacturing as a key to competitive advantage in one particular product market. Ultimately it placed equal if not greater weight on the medium- and long-term interests of the multinational group world-wide than on the short-term profit maximization of the UK operating company and its general wiring division. What emerged was essentially a technology-driven strategy.

## Designing and modifying a technology-driven strategy

The idea of technology-driven strategy came to prominence in the USA in the early 1980s. This was a time when new computing and information technologies were beginning to come on stream and business leaders were becoming increasingly aware that they were failing to meet the competitive challenge from Japan and other Far East countries (see Chapter 2). By 1983 the idea had become a key element in a full-blown critique of US corporate governance, with its apparent fixation on short-term financial objectives and its lack of preparedness to engage in long-term product and market innovation (see Abernathy et al., 1983). The success of Japanese companies was seen as arising directly from their strengths in technological innovation and the ability this gave them to develop innovative product designs and new systems of production and management (Pettigrew and Whipp, 1991: 17, 18).

Such a strategy was adopted by Pirelli General in 1984–5. In the previous decade the company's general wiring factory had been languishing under low levels of investment and technical innovation. The Aberdare project was driven by senior managers at group, company and divisional level, all of whom came from an engineering background. They were convinced that, in the medium to long term, the company would only be able to maintain competitive advantage in the general wiring product market by investing in state-of-the-art computer-integrated manufacturing technology. This, it was believed, would enable them to reduce the costs of labour, materials, work-in-progress and finished goods stock, and at the same time adapt flexibly to changing customer requirements. In addition, being a multi-national company with a dozen general wiring factories all over the world, senior multinational managers saw an advantage for the group world-wide in carrying out an experiment in one national operating company. They therefore invested some of the group's research and development budget to help one of its national operating companies experiment in areas of technical innovation – areas in which it would not have invested if the decision had been made on purely commercial, national company grounds.

The extent to which this was a technology-driven strategy became clear between June and September 1985, when calculations showed that the projected new factory would have a spare capacity of around 30 per cent if it produced – as originally planned – roughly the same quantity of cable as the existing general wiring factory. A revised proposal was then drawn up which assumed a greater use of capacity and increased market share. These assumptions were not based on detailed market research, but on the need to justify the investment in a computer-integrated manufacturing system and make it operate more efficiently.

This technology-driven strategy was not simply one element of a wider business strategy. It dominated the whole approach of managers at sector, national operating company and divisional level in the first years of the factory's operation. From 1987–9 short-term financial and commercial considerations were seen as less important than engineering and technical experiments on behalf of the group. However, between 1989 and 1991 new senior executives were appointed at group and UK company level who were less committed to the experimental nature of the factory and more concerned to see it perform as a conventional commercial and production operation. With the onset of recession in the UK construction industry, a technology-driven strategy based on a commitment to produce at or near full capacity and increase market share was felt no longer to be a viable option.

The new strategy that emerged in the early 1990s still built on the competitive advantages flowing from computer-integrated production and management systems. Indeed, between 1991 and 1993 the retreat from full automation was reversed and incremental advances were made in achieving a more integrated computerized system of manufacturing and logistics management. In this sense the original 'technology strategy' remained intact. However, the overall business strategy was no longer driven mainly by technology, but by commercial considerations – above all by a re-analysis of customer requirements and a reorientation of production towards those products which would provide the best overall level of profitability. In this context, a rethink of commercial strategy linked with tighter financial controls, staff reductions, and the pursuit of efficiency improvements across the board, became the order of the day.

### Greenfield or brownfield site?

As we saw in Chapter 3, the conventional wisdom of US research, based on the experience of the 1970s and 1980s, is that it is easier to innovate successfully on greenfield sites because 'the spatial design, technology and people involved are all starting afresh' (Whyte, 1990: 337). On a greenfield site, too, these can all be meshed together into a 'congruent total system' (Lawler, 1978: 6). Recent empirical UK research, in contrast, has suggested that organizations contemplating major organizational change should not be so concerned with the choice between greenfield or brownfield sites as with the adoption of a 'greenfield philosophy' (Newell, 1991). In support of this argument there are some well-documented examples of companies that have achieved successful technical and human resource change on brownfield sites by harnessing the already existing goodwill and cooperation of a highly-skilled and committed workforce. Such examples warn against too easy an assumption that greenfield sites will always offer a better basis for implementing advanced manufacturing systems and other organizational innovations.

However, in the case of the Aberdare project, the greenfield option did provide a highly promising 'opportunity structure' to facilitate the establishment of a state-of-the-art computer-integrated manufacturing plant. The factory layout and the machines in the company's existing brownfield UK factories (at Southampton and Aberdare) were relatively inflexible, labour intensive and costly in terms of heating and maintenance. Achieving major human resource change in these plants would also have come up against hierarchically-oriented line managers and workforces rooted in formalized working practices with clear job demarcations and divided occupational and union affiliations. In contrast the greenfield site at Aberdare created the opportunity to design a highly flexible and production-efficient factory on modular and zonal principles. It also allowed the company to negotiate a comprehensive single-union agreement, recruit a new workforce, and establish a new organization culture and new patterns of flexible team working. Without the *tabula rasa* of a greenfield site it is highly unlikely that the intended degree of technical and human resource innovation could have been achieved.

However, the greenfield site location did have one major unanticipated negative effect. The site was chosen partly for financial reasons, such as government grants and the cheapness and availability of labour. However, it was also meant to isolate the new factory geographically from the influence of the more traditional working practices and union arrangements in the company's other factories. This, in turn, proved to be a major barrier to the ability of the company to recruit its best middle and senior managers to take up positions there. For many, relocation to an area where the standard of living and general quality of life were perceived to be much lower than in the South of England was too big a price to pay for themselves and their families. In Chapter 6 we noted that one of the main features which distinguishes human resource management from more traditional forms of personnel management is the quality of management and management practice. The choice of Aberdare worked against the recruitment of the company's best staff to manage the new factory.

## Implementing strategic change

The implementation of strategic change is a 'cascade process' of ever more detailed decisions and policy developments which occur in the context of primary strategic aims (see Hrebeniak and Joyce, 1984). In Chapters 5 and 7 we identified a range of factors associated with successful implementation (see Child, 1984: 291–2). In particular we noted the importance of the support of top management and the need to establish mechanisms to monitor and evaluate change and its outcomes systematically. We also looked at research on the advantages and disadvantages of different methods or styles of implementing strategic change. The general conclusion was that the choice of a particular approach was likely to be contingent on the seriousness of the problem with which the change was intended to deal and the anticipated degree of resistance to change within different sections of the workforce. All things being equal, a participative approach in which senior managers take employees into their confidence about the reasons for change and consult them about ways to achieve their objectives was found to promise a better chance of smooth implementation. In the longer term, too, this approach is also able to provide a better basis for developing what Child has called an 'adaptive learning capacity' (1984: 285).

The story of human resource and technical innovation and change at Pirelli Aberdare provides a classic illustration of strategy as an emergent property. Strategy evolved not only from the process of setting primary strategic aims (Chapters 1 and 2), but also from the detailed elaboration of these aims (Chapters 3, 4, 6 and 8) and their eventual translation into practice (Chapters, 5, 7, 8, 9 and 10). Events also provided an all-too-familiar confirmation of US research on the most frequent implementation problems encountered by companies engaged in managing strategic change.

Some of the problems could be attributed to project management, for example relations with equipment suppliers and the lack of coordination between the implementation of technical and human resource change. Perhaps more importantly, no mechanisms were put in place during the first two years of factory operation (1987–9) to monitor the process and outcomes of change systematically. Neither the performance of the project manager (David Regan), nor that of the project as a whole, were subject to a detailed review. This can be explained in part by the status of the project as an experiment on behalf of the group world-wide in which the company's normal financial and managerial rules did not fully apply. But the outcome was what one manager called an 'extended honeymoon' period in which no one was really called to account to give clear direction to the project when it was in obvious difficulties.

The first steps to carry out a more systematic review of the project were taken in January 1989 with the secondment of a senior corporate manager to investigate the performance problems and make recommendations. But it was only when the new managing director instituted a system of monthly operations review meetings on site in the summer of 1989 that the senior management team, and the main functional line managers, were regularly and directly called to account for their performance. It was at this time that the technology-driven strategy began to be modified and a new business strategy evolved, initially giving priority to production performance and output, subsequently focused more on profitability, customer service, continuous improvement, and a revised commercial strategy.

## Human resource management and single-union agreements

In Chapter 6 traditional personnel management and human resource management were contrasted as different approaches to the management of people at work. For some, the essential difference between the two lies in whether or not organizations take a strategic view of personnel or human resource issues. For others, it is not the existence or absence of a strategy which is crucial, but the direction and objectives underlying strategy (see Guest, 1989). For Guest the distinctiveness of HRM can be encapsulated in the proposition that there is an organizational payoff for companies if they develop a combination of policies designed to meet four key human resource objectives: organizational integration, employee flexibility, high commitment, and high quality (Guest, 1989: 42).

In the UK, the development of comprehensive human resource strategies has been strongly associated with companies setting up plants on greenfield sites, in many cases with comprehensive single-union agreements (see Chapter 8). The particular form of HRM adopted has been strongly influenced by the ownership and nationality of the company concerned. American companies, coming from a traditional HRM culture of anti-unionism and individualism, tend to pursue a non-union path; Japanese companies, with their tradition of 'house' unions, are likely to seek some kind of non-union company council or a comprehensive single-union agreement; European companies, more accustomed to pluralist systems of labour management, tend to opt for single-union recognition and some form of collective bargaining arrangement alongside more individualistic HRM policies.

The story told in this book involves a conscious attempt in one division of a multinational company to move from a traditional style of personnel management – as practised at the old Aberdare site and the Southampton general wiring factory – to a new and comprehensive European-style HRM strategy. Human resource issues such as 'headcount', teamwork and single-union agreements were present in early discussions about setting up the new factory. However, they tended to be secondary to technological and financial considerations, particularly in the build-up to the board decisions of June and September 1985. It was not until 1986 that the project manager and a group of three personnel specialists set about designing a coherent and comprehensive human resource strategy which would be appropriate to the business and technical requirements of the new plant (organizational integration) and, where possible, would be consistent with progressive personnel practice.

Full employee flexibility – in particular task flexibility and the breakdown of demarcations between production, maintenance and office staff – was seen by senior managers as a central operational and human resource requirement of the new plant. Together with a participative management style and self-supervision, it was the cornerstone of the new HR strategy, and one from which most of the other policies – for example on single status, skill-based pay and single-union recognition – followed. And yet, within three years, the company retreated explicitly from the requirement for full flexibility of the workforce, not only because it proved to be operationally unnecessary, but because its full implementation was perceived to be contrary to some of its other key human resource objectives. For example, the site's skill-based pay system, which created a financial incentive for individual employees to acquire skills and become fully flexible as fast as possible, came into direct conflict with the need for thorough training, skilled working in particular areas, and team spirit and cooperation. Despite the retreat from full flexibility, however, the level of flexibility achieved in the production area at Aberdare – with most producers trained and able to work competently in five highly complex machine areas – was substantially greater than in any other Pirelli factory in the UK and in UK companies more generally.

As for the single-union agreement, the company achieved most if not all that it had intended when it decided to go down this route in 1986, including the end of multi-unionism and the avoidance of work stoppages. The non-management workforce's experience of the agreement was much more contradictory. By the end of 1992, during which there had been an extended pay freeze and a programme of job cuts, the vast majority – as individuals and as union members – felt a strong sense of *powerlessness in industrial relations* matters such as pay and staffing levels. A substantial minority had begun to question fundamentally the future role of the union on site (see Chapter 8). This was in stark contrast to the overwhelmingly positive feeling of *empowerment in work relations* through the implementation of self-supervision and the establishment of quality improvement groups (see Chapters 7 and 9).

## Total quality management

The idea of total quality management was first implemented systematically in post-war Japan (see Chapter 9). It grew out of the application of rigorous quantitative performance techniques – such as statistical process control (SPC) – to manufacturing processes, and was then developed into a fully-fledged holistic management theory (see Hill, 1991). For its leading proponents, such as Edward Deming (1982, 1986), TQM means shifting the emphasis from inspection for quality after the event – usually via a separate quality control or customer service department – to preventing defects or poor service through the incorporation of quality assurance into the day-to-day responsibilities of all staff. The holy grail of TQM in manufacturing organizations is to create a virtuous circle in which the elimination of waste leads to improved quality, lower costs, greater efficiency and competitiveness, and long-term profit maximization. As Hill points out

(1991: 403), short-term profit maximization is alien to most theories of TQM.

Between 1987 and 1990 David Regan and the senior management team at Pirelli Aberdare created an organization culture – reflected in the practice of self-supervision – in which all members of staff were expected to take responsibility for the quality of their work. In interviews conducted in 1990, the non-management workforce expressed overwhelming support for self-supervision and was strongly committed to involvement in the day-to-day management of product quality. But the introduction of TQM in 1991 translated this general 'cultural' commitment into a series of clearly-defined and measurable projects and initiatives for quality improvement. What had existed prior to TQM was what one manager called 'TQM without measurement' – quality management without clear priorities, targets and deadlines.

The first stage of the TQM programme at Aberdare comprised three 'fast track' projects concentrating on problems identified by managers as priority areas for improvement. The project teams were composed of a mix of managers, systems engineers, process technicians and producers, and each achieved tangible improvements in production efficiency and information management within a six-month period. But these projects had three drawbacks: they were time-limited, they only dealt with managerially-defined problems, and they only involved a small number of employees.

From March 1992, projects were set up in all areas of the factory in the form of quality improvement groups (QIGs), teams of employees explicitly authorized by management to spend part of their working week improving their own work process. From mid-1992 SPC techniques were introduced, under the control of the site quality assurance manager, to investigate ways of improving performance and materials management in the metallurgy area. One of the main techniques used to involve all employees in TQM at Aberdare was the PICA board, a cause and effect diagram displayed in the relevant work area on which all staff were invited to write or attach notes identifying problems, suggestions or comments.

In 1993 the QIG in the 'packing and assembly' area received 88 suggestions from staff, of which 63 were completed during the year. Visits by the group to a material supplier led to changes in material use and the redesign of cardboard cartons, which resulted in an improvement of output by 4 per cent per shift. Both initiatives had originated in the frustration of producers at materials which had jammed while passing through the packaging process. Both resulted in significant time and money savings for the company and increased job satisfaction for the workforce.

The immediate aim of TQM at Pirelli Aberdare had been to achieve certain quantifiable targets by the use of methodologies and techniques such as SPC, QIGs and PICA boards. But the longer-term result – part intended, part unintended – was as much processual as substantive. By 1993 it had helped to foster an organization culture in which the idea of continuous improvement of all work processes by the people actually involved in them had become institutionalized. While TQM clearly served the interests of management, it also created a sense of involvement and empowerment among the workforce which was far more deep-seated and meaningful for them than involvement in more formal consultative meetings such as team briefings, business review meetings and general meetings.

## *Leadership*

In Chapter 10 we presented a three-pronged definition of leadership adapted from the work of Peter Drucker (1955) and John Adair (1990): giving direction, raising performance, and building personality and teams beyond their existing limitations. Leadership was distinguished from management, with its more technical preoccupation with the achievement of specified goals (see Bryman, 1986: 6). It was noted that, according to US research, the most effective leaders were those who got the job done and maintained good team relationships, that is, married 'task-oriented' and 'employee-oriented' leadership (see Fleishman, 1953a, 1953b).

Initial research on leadership was dominated by the *trait* approach, the search for measurable characteristics of effective and successful individuals. More recent research has placed a greater emphasis on the *context* within which leaders operate in order to explain why certain types of leadership behaviour are successful and others not. From this we hypothesized that some types of leadership behaviour and style may be more appropriate and effective at certain stages in the process of strategic change than at others.

In the Aberdare story there are many examples of people – both as individuals and members of teams – who exercised different kinds of leadership at different times on different issues and at different levels. The project was decisively shaped at the outset by the vision, commitment and support of cable sector general manager, Arthur Stokes. The influence of his 'tablets of stone' speech is clear proof of Kanter's contention that 'if there is one domain over which top executives have control it is organization culture . . .' (1983: 362). But perhaps the most interesting contrast in leadership behaviour was provided by four individual members of the Aberdare workforce: David Regan, project and divisional manager (1985–91), Michael Elliott, operations manager and then Regan's successor as site manager (1990–2), Martin Parker, head of maintenance (1991–2), and Stuart Wood, systems engineer in charge of POMS (1990–3).

Regan personified the ideal of the high-technology, high involvement computer-integrated factory throughout his almost seven years of direct involvement in the project. His greatest talent was to enthuse, motivate and gain the commitment of the people who worked with him and for him to achieve this ideal. His was a charismatic, inspirational, employee-centred leadership which generated enormous mutual loyalty between him and his staff. He played a crucial role in getting the project off the ground and sustaining it in its most difficult period. However, his strengths were also weaknesses. He established few rules. He trusted management and non-management staff alike to be self-supervisory. He did not create an environment in which specific targets were set for individuals, teams and the factory as a whole, and in which their performance was monitored systematically.

This kind of leadership was eventually provided by Michael Elliott when he took over as site manager in September 1991. Elliott acted as a catalyst for change, building on the creativity and commitment of the workforce generated during Regan's time, but channelling it in a more focused way. He set priorities and used the TQM programme to set measurable and achievable targets for individuals and teams. It was a more task-centred leadership than Regan's, but one which also worked through senior and junior managers and bound them into the general effort. Elliott also facilitated and encouraged two different kinds of lower level task-oriented leadership, in two specific areas of the factory: maintenance and systems engineering.

In 1991 Martin Parker took over as manager of the maintenance department. At that time it was in some disarray. Within 12 months he had given it a direction and a sense of purpose. He established tight budgetary controls and a new workload management system; introduced the innovative and highly-effective technique of predictive maintenance; began to delegate much of the department's routine work to producers; and persuaded Elliott to reduce the number of maintainers by a third and establish a more focused role for them as a highly skilled specialist support team. Parker was not generally popular with his staff. However, most admitted that his almost completely task-centred leadership had improved the efficiency and general contribution of the department beyond recognition. He had given it both leadership and management.

Stuart Wood exercised a very different kind of task-centred leadership between 1990 and 1993. He derived his leadership potential not from positional or line management power, but by his knowledge of computerized management systems, his commitment to working with managers and staff to improve them, and his willingness to put in long hours to achieve the desired ends. As with Parker his personal characteristics suited the context. When he was put in charge of the development of the plant operations management system (POMS), the external consultants who had designed the basic software were just ending their involvement and he was able to have the kind of free hand which his predecessor had not enjoyed. Elliott had just been appointed operations manager and was keen to listen to Wood's ideas and persuade other managers to adopt them. At the same time, the company was about to introduce its TQM programme, and the fast track project in which Wood was subsequently involved as team leader demonstrated the benefits of computerized management systems when professional engineers, production managers, process technicians and producers pooled their knowledge and experience. Between 1990 and 1993 Wood played a pivotal role in extending the computerization of the factory and securing acceptance for POMS as the central mechanism through which most aspects of the site were managed. Wood was not a leader of the 'grand gesture', more a 'quiet innovator' (Kanter, 1983: 354). Unlike some of the other leaders, by 1993 he was quite happy to return to the ranks.

## Managing Strategic Innovation in the 1990s

In considering the more general implications of the Pirelli Aberdare story, what is immediately apparent is the extraordinarily turbulent internal and external environments with which managers and staff at all levels had to cope, while at the same time engaging in the design and implementation of a large-scale and highly innovative experiment.

In 1984, one senior multinational group manager had come to the conclusion that the threat of competition from low labour-cost countries in one of the group's core product areas – general wiring cables – was potentially so great that the group would have to find new ways of competing if they were to survive in this particular market. Just at this time, new computer technologies were becoming available which opened up the possibility for technically advanced and comparatively cash-rich companies such as Pirelli to automate production, reduce the costs of labour and stocks, and at the same time respond speedily and flexibly to customer orders and changing customer demands. Putting these two things together, Stokes concluded that it would be to the group's advantage to conduct an experiment in one national operating company to see whether the potential of computer-integrated manufacturing could be converted into reality. He saw it as a project which would be unlikely to be commercially successful in the short term, but which could well put both the national operating company and the group in a stronger and more profitable position in the market-place in the medium to long term. At the same time, the experiment would be valuable in itself, as the group would be able to learn both from its successes and its failures.

When this strategy was developed and the decision taken to go ahead with the experiment in 1984–5, it would have been almost impossible to predict the breadth and depth of the political and economic changes that would take place in the following 10 years: the stock market crash of 1987, the collapse of communist regimes in Eastern Europe, the end of the cold war, the breaking down of tariff barriers world-wide (and particularly in the European Union), the exponential growth in global competition, the increasing pervasiveness of competition and markets in all aspects of public life, the volatility of exchange rates, and finally, in the early 1990s, the longest international recession since the Second World War. Against such a background it is questionable whether any organization would have been able to implement a highly innovative and experimental strategy without significant modifications and adjustments.

The internal environment within which the experiment was conducted was also highly turbulent. In 1989, the Aberdare factory was a semi-autonomous, profit-accountable division; it had around 220 employees, including 50 sales and distribution staff based in five geographically dispersed distribution depots; it also had a senior management team including an operations manager, divisional accountant, commercial manager, and personnel and training manager. Just four years later, in 1993, the factory was but one part of a larger division; the core workforce was less than 130 strong; over 20 of these staff were working in a new automated warehouse whose opening had led to the closure of all but one of the remaining five distribution depots; all the plant's needs for plastic sheathing material were now supplied on site by a new PVC factory; all the senior management positions at the site had been abolished; all but one of the postholders in 1989 had now left the company; and the managing director had also changed twice. At multinational group level, the number of factories world-wide had been reduced from 102 to 80 between 1992 and 1994. Over the same period the group's total workforce had been cut from 50,000 to 40,000, and 600 managers had changed positions.

Managing an organization, let alone managing a large-scale technical innovation, against the background of such a turbulent internal and external environment, is difficult enough. But what became abundantly clear as the project progressed was that what may have appeared at the outset to be a major innovation in one aspect of the organization – technology – proved to have major implications for virtually all other aspects of the organization too. For example, the high degree of computer-integrated production at Aberdare would not have been possible without the innovations which were made – incrementally – in site location, building and factory layout. The same is true for the innovations – some planned from the beginning, others elaborated more incrementally as they went along – in organization culture, organization design, and human resource management. One of the most important 'lessons' of the Aberdare story is the extent to which technical, production, distribution, human resource, financial and commercial activities are interrelated. Ensuring their conceptual and practical 'fit' was a highly complex and never-ending task, crucial to the realization of wider corporate, business and operational strategies.

To what extent can the Aberdare experiment be regarded a success? The answer to this question depends critically on the criteria and the timescale chosen to measure 'success'. The project began life as an experiment in one division of one national operating company on behalf of a large multinational group. The corporate and commercial strategy chosen would not have been adopted if the goal had been to achieve the best short- and medium-term return on capital invested in one division of one national operating company. In the latter context the Pirelli group's more conventional Spanish building wires plant, built at around the same time as the Aberdare factory, represents a clear alternative 'national company' model (see Chapter 5).

By 1989, a new senior management team at multinational group and UK operating company level began to play down the experimental function of the Aberdare factory on behalf of the group and to concentrate on making it a commercially viable, self-standing operational unit in its own right. However, at exactly the time when it would have been able to show its full potential as a self-contained operational unit (in 1990–1), the UK construction market entered the most extended recession since 1945, from which at the time of going to press (1995) it has not fully recovered.

If the factory is to be judged by the 'bottom line' criterion of profits before tax, then its performance between 1987 and 1993 cannot be counted a success. In contrast, by 1993 it did prove to be a success in terms of production efficiency measures (reduced value of raw materials stock, work-in-progress, scrap levels, finished goods stock), distribution costs (with its new automated warehouse on site replacing the eight geographically dispersed distribution depots of 1985) and customer satisfaction (99 per cent satisfaction rate, with guaranteed 24-hour delivery anywhere in Britain). In personnel terms, too, labour costs had been more than halved compared with the old factory, there was a highly cooperative and relatively well-paid workforce with ever increasing skills (e.g. the development of shopfloor production workers into college-trained production engineers, the creation of opportunities for producers to become producer/maintainers). No days had been lost through strikes since the factory opened and absenteeism was at an all time low. Taking all these things together, the factory was clearly in a strong position to compete successfully in a highly competitive UK and European market-place and to continue to provide satisfying and well-paid jobs for its reduced workforce.

In the 10 years covered by this book, Pirelli Cables as a company underwent a number of major changes: in national and international structure, in workforce size, and, above all, in general business orientation. Its strong commitment to engineering excellence and production performance was retained, but now tempered by a much greater emphasis on customer and market requirements, a keener focus on quality, and much greater awareness of the extent to which future competitiveness depended on performance in European and world markets. The Aberdare factory was subject to these changes in the same way as other factories in the company. But what marked out the Aberdare experiment from the rest of the company, and from much of British industry, was its innovative experiment in how to organize work to ensure a practical commitment to continuous improvement. It can be summed up in one word: teamwork.

In 1993, executives from 25 leading UK companies – chaired by the chief executive of IBM (UK) – met under the auspices of the Royal Society of Arts to try to identify the key characteristics of what they called 'Tomorrow's Company'. One of the central findings of their final report was the 'strong correlation between the companies with the highest emphasis on individualism and the poorest economic performance over the last 10 years' (RSA, 1995). This was in contrast to the high levels of commercial success of companies adopting a team approach, particularly through collaboration between managers and employees, but also between companies and their investors, suppliers, customers, and local communities. It was in this context that the report cited the telling words of the Japanese industrialist Konoke Matsushita, spoken prophetically in 1978 to a group of American business leaders:

> We are going to win and the industrialized West is going to lose out . . . the reasons for your failure are within yourselves. With your bosses doing the thinking while the workers wield the screwdrivers, you are convinced deep down that this is the right way to do business. For you the essence of management is getting the ideas out of the heads of the bosses and into the hands of labour . . . For us the core of management is the art of mobilising and putting together the intellectual resources of all employees in the service of the firm. (RSA, 1995)

Whatever its strengths and weaknesses, the Pirelli Aberdare experiment put such an approach into practice between 1987 and 1993 – and showed that it could work in the UK with a highly traditional workforce.

In today's continually changing competitive environment, there can be no happy endings. It is only the most foolhardy commentator who is prepared to go into print and identify particular companies or plants as excellent. Just a few years after the publication of Peters and Waterman's *In Search of Excellence* (1982), which celebrated the US's most successful companies, a number of them were in crisis and some were making huge losses. The history of the IBM in the early part of the 1990s is a salutary example of a successful 'blue chip' company falling on hard times, and then beginning to regenerate itself. What is clear about the Aberdare experiment, however, is that between 1987 and 1993 the management and workforce created an environment which had many of the features of what Kanter described over a decade ago as the commercial organization of the future:

> . . . we need to create conditions, even inside larger organizations, that make it possible for individuals to get the power to experiment, to create, to develop, to test – to innovate. Whereas short-term productivity can be affected by purely mechanical systems, innovation requires intellectual effort. And that, in turn, means *people*. All people. On all fronts. In the finance department, the purchasing department, and the secretarial pool as well as the R & D group. People at all levels, including ordinary people at the grass roots and middle managers at the heads of departments, can contribute to solving organizational problems, to inventing new methods or pieces of strategies. (Kanter, 1983: 23)

The future success of Pirelli Aberdare will depend crucially on its ability to sustain such an involving, innovation-led environment and to continue to develop the adaptive learning capacity of its workforce.

# 13

# Postscript: The Aberdare Building Wire Project – A Personal View

*Stanley Crooks, former Head of Cable Sector, Pirelli Group*

## The Strategy

It is not often that an industrialist finds one of his strategies subjected to a scrutiny as searching as the one which Jon Clark recounts in this book. What Jon has achieved, through tracing the Aberdare project from conception to fruition, is a critical analysis of how practice in this case compared with both intention and theory. In so doing, he shows a perceptive understanding of the crucial influence of human characteristics, whether weaknesses which impose constraints or strengths, sometimes from unsuspected sources, which provide impetus or inspiration at critical moments in a project's development. He has also, I believe, demonstrated rather clearly that (as he puts it in Chapter 5):

> In practice, strategies are not fixed at some single decision point; they are evolved and elaborated continually as the internal and external environments of organizations change.

Jon asked me, after I had read the first draft of his book, how I, the initiator, felt about the way the project had gone, and whether in retrospect I felt that it was a sound decision.

Well, hindsight is not a particularly useful management tool. For one thing, 'strategies evolve', just as Jon says. For another, while one can trace the development of a project from conception to fruition, and can describe the choices which were made between alternative courses of action at specific stages in the process, it is impossible to trace what would have happened (or guess what might have happened) had different choices been made.

Furthermore, one of the crucial elements of internal change which influence the way in which strategies evolve is the change of key people in the organization during the course of that evolution, and there is always a temptation (which I believe it wise to resist) for an interested party to say, 'After all, I would not have done it that way'.

The decision which I took in 1984, as General Manager of the Pirelli Cable Sector, was to conduct a research project into a business process. We were satisfied that our R & D programme was successfully creating competitive advantage and good profits for us in high technology areas such as optical fibres, very high voltage power cables and submarine cables, but we were not well pleased with the economic performance of our 'bread and butter' products. In these low technology areas, product designs were rigidly standardized, and there was little scope for conventional product R & D. The traditional approach of periodically modernizing factories, although it had to be done to defend our market position, gave little longer-term competitive benefit because the latest designs of machinery were equally available to our competitors. We were seeing that, in our markets in developed countries, our businesses in these low technology product segments – always subject to strong competition from domestic manufacturers – were beginning to suffer increasingly from imports from countries with low or subsidized costs.

It would, of course, have been possible to decide to run down, sooner or later, these business segments in developed countries. This is exactly what many European and American companies had been doing for quite some time in many low technology business areas. (It is only in the late 1980s and the 1990s that we have begun to notice that the decline in the manufacturing base has rather unpleasant consequences for employment and the balance of payments – but this is hindsight, isn't it!) In our case, we believed that we must take account of an important fact of a cablemaker's life, that our high technology products, although profitable, follow long business cycles or are subject to highly fluctuating demand, and they call for heavy investments in R & D and in equipment. One prepares five-year, not one-year, budgets for supertension cables, and even longer ones for submarine cables: inspired *fore*sight is a *very* useful management tool! So, a good chunk of low technology business provides a base load and flexibility.

However, no product can be said to be strategically important to an organization if it consistently makes inadequate profits. We had to consider whether an alternative way could be found of doing this sort of business so as to give us a competitive advantage. We had been impressed by recent progress made by the computer industry in developing small personal computers which could be dispersed around an operating site and networked together so as to give flexibility in use without loss of computing power. (Remember that it was 1984.) It seemed to us worth while to investigate, purposefully, the extent to which such systems could be used to reduce employment costs, scrap, work-in-progress and finished goods inventory and so – exploiting local market presence by providing quick and responsive customer service with low working capital – counter the low employment cost advantages of importers.

We recognized that unproven techniques would have to be explored, both in improving mechanical solutions for standard operations to make them suitable for computer control, and in designing new software (though the computer hardware was not seen as a problem). We therefore knew that significant up-front R & D costs would have to be incurred in investigations and pilot trials, and that a measure of excess capital investment would be needed to encourage equipment manufacturers to develop innovative mechanical solutions. The extent of these items could not be accurately evaluated in advance and so represented a financial risk. Finally, we realized that a project like this could only be effective if applied to a complete operating unit, and that implementation time between approval and achievement of planned operation at regime would be longer than for a conventional project (that is, in excess of the normal 3–4 years) and that, in the event of delay in commissioning, back-up production facilities would be needed to avoid loss of market share. The cost of providing this back-up – for a period which could not be determined in advance with any certainty – was a further component of financial risk. For all these reasons, we decided that it would be sensible to confine the risks to a single trial project for the whole Cable Sector, and apply new solutions elsewhere only when thoroughly proven.

## Specific Objectives and Achievements

Jon Clark has described how we went on from there, selecting the UK as the country, Aberdare as the site, and so on. To put the project in perspective it may be noted that, in 1985, the UK building wire business was just over 1.5 per cent of total Cable Sector turnover and just over 10 per cent of total Sector building wire sales. We were making building wires in 11 countries, of which 4 could be classified as low employment-cost countries. Perhaps it does not come out clearly enough in Jon's account that Southampton was not really a viable alternative: all of the buildings on the site, on reclaimed land, needed replacing – not just those occupied by the building wires unit, which was one of three major production units on the site.

When, in 1985, we set the operational targets which we expected Aberdare to achieve when operating at regime, we set them for 1990: five years ahead. (At about the same time, we approved a conventional modernization of our Spanish building wire business – a new factory – which successfully achieved regime operation in the normal three years.) By 1990, Aberdare had achieved some of its targets – those for space occupation and personnel numbers. But 1990 turned out to be the first year of the UK's long recession, and the volume, productivity, scrap and working capital targets took a further three years to achieve. However, by 1993, productivity was three times the 1985 level, and 50 per cent better than the original 1990 target. The profitability target was not met until 1994 – in 1993 prices were still low, the recession having brought the usual erosion of prices as manufacturers chased market share in the face of a rising tide of imports.

## The Lessons

However, the recession was not the only reason for late achievement of targets. Some lessons can be learned. Identifying what those lessons are is, inevitably, a matter of subjective judgement. Early in the project, some mechanical problems were encountered which had nothing to do with the sophistication of the project, but nevertheless caused delay while they were rectified by a machinery supplier: choice of a different supplier (at higher cost) would probably have avoided that particular problem. Some of the innovative mechanical solutions proved more elusive than had been hoped, and one quite important one had to be abandoned in favour of a more conventional solution: it may be argued that a delay caused by taking more time for trials in the development workshop might have been shorter than the delays in production. Software development, as in so many CIM projects around the world, proved to be even more difficult than already expected. The fact that the software contractor was based in Italy certainly caused some communication problems at the critical phase of refining the design concepts to match the realities of the shopfloor; whether the 1990 decision to 'reculer pour mieux sauter' led to a more rapid result than persistence would have done can never be proved. Changes of personnel – some planned for the greater good of the company as a whole rather than for the benefit of the project, some unplanned departures – clearly caused discontinuity and delay, and would probably not have occurred to the same extent or have had such an effect in a shorter project of normal timespan with more conventional problems to solve.

## Reprise

Would I, knowing what I know now, take the same decision again? Yes. The threats foreseen were real. Imports rose from less than 10 per cent in 1984 to over 30 per cent in 1994. Aberdare is currently competitive at the prices which can be achieved in the market-place.

I find it hard to agree with Jon Clark's view that the strategy was 'technology-driven' – as opposed to 'commercially-driven'. I would prefer to say that we were seeking a commercial solution using technology as a tool. The technology of computer integration, in embryo in 1984, works – and works well – in 1994. The effort has been worth while.

I believe there is more to be squeezed out of it – the principle of continuous improvement in performance applies just as strongly to a sophisticated business system as to a simple one (my successors understand this very well, and will not let up in their efforts in that direction). There may still be something more to be done in extending the integration beyond CIM to CIB, although some of the limits of realistically useful extension have already been tested.

I confirm then, that if I were back in 1984, I would launch the research project again. I would also approve the *conventional* project for Spain again. The Aberdare solutions *are* now proven and ready for broader application and, if I were still 'on active service', and had to approve a complete new plant now, I would expect to be presented with proposals showing the costs and benefits of the latest versions of *both* solutions before making a decision on which to go for.

# Technical Appendix: Research Methods

The research on which this book is based was designed as a longitudinal study of the process and outcomes of human resource and technical innovation. It was longitudinal in two senses. First, it aimed to tell the story of Pirelli's new Aberdare factory from early 1984 – the year in which the idea of an experimental factory was first mooted in the company – to the end of 1992, when the primary research would be completed. Much of this research would of necessity be retrospective. Second, the intention was to carry out two periods of intensive fieldwork in the factory – the first in mid-1990, the second in mid-1992 – to collect primary data about changes and continuities in the experience of managers and staff working at Aberdare over these two years.

The general mix of methods chosen – 45-minute semi-structured interviews with all non-management staff; extended, open-ended interviews with all managers; predistributed personal questionnaires for all interviewees requesting basic data about their personal characteristics, work and education/training histories; observation of non-management work; background interviews with senior company managers; internal documentary sources such as company handbooks, newsletters and circulars – had been used successfully in a previous project on the modernization of telephone exchanges in British Telecom (see Clark et al., 1988: 224–8).

However, there are a number of differences between the mix and relative importance of various methods on the two projects. The work tasks of telecom maintenance engineers were comparatively easy to classify and thus to record in daily work diaries. In contrast, work tasks were not so easily classifiable at Aberdare, and so work diaries were not used. In contrast, background interviews with senior company managers – some 20 altogether, many lasting up to five hours over two meetings – have figured more prominently as a method of primary data collection in the Aberdare study, and played a central role in the reconstruction of the initial years of the project and of the complex interplay between multinational, national and site management.

## Interviews

When I first conceived the idea of the project in May 1989 after a first visit to Aberdare, I intended to interview all managers and specialists (of whom there were 35 in mid-1990), and a sample of non-management staff (of whom there were 112 at the time). However, when I secured research funding – both from the Leverhulme Foundation in 1990 and subsequently from the ESRC-funded Industrial Relations Research Unit at Warwick University from 1991–3 – and when it emerged that I would be able to have almost total access to managers and staff at Aberdare, I decided to go for broke and interview the whole workforce.

The non-management interviews were carried out in two periods of five weeks (August and September 1990 and 1992). I would arrive at Aberdare by Monday lunchtime, stay overnight from Monday through Thursday, and travel back home

**Table A.1** *Number of Aberdare staff interviewed by function and level, 1990 and 1992*

| Interviewees by function and level | 1990 | 1992 |
|---|---|---|
| Administrators | 19 out of 19 | 16 out of 17 |
| Maintainers | 13 out of 14 | 11 out of 11 |
| Producers | 71 out of 78 | 67 out of 69 |
| Warehouse staff | – | 14 out of 14 |
| Systems engineers | 9 out of 10 | 5 out of 5 |
| Other senior management staff | 25 out of 25 | 22 out of 22 |
| Total | 137 out of 146 | 135 out of 138 |

early Friday afternoon. During the period of four full days each week, I would interview between five and ten staff each day for between 30 and 45 minutes. Without the assistance of personnel manager Bob Nicholas and two personnel administrators, Debbie Woods and Karin Rees – who organized the programme of interviews and ensured that staff had filled out their personal questionnaires and knew when and where they had to appear – it would not have been possible to complete such an exhausting schedule.

In the end, I was able to interview 94 per cent of the total workforce in 1990 and 98 per cent in 1992 (see Table A.1). No one refused to be interviewed, those who were not interviewed were either absent long-term sick or scheduled for the end of the interview cycle and unable to make the appointment on the day for one reason or another (sickness, holiday, work requirements). All managers and specialists on site were interviewed using a more open-ended interview schedule. These interviews varied in length from 45 to 90 minutes and took place in July and August 1990 and 1992, just prior to the main interview programme.

A number of characteristics of the Aberdare workforce – gleaned from their completed personal questionnaires – are worthy of note. First, unlike companies such as Nissan, which under similar circumstances recruited a comparatively young non-management workforce to operate its new manufacturing plant in Sunderland (see Wickens, 1987), Pirelli General had a 'balanced' non-management workforce in terms of age range (see Table A.2). Pirelli also chose consciously to recruit a number

**Table A.2** *Age structure of the non-management workforce, 1990 and 1992 (%)*

| Age | 1990 (n = 103) | 1992 (n = 108) |
|---|---|---|
| 16–24 | 18 | 13 |
| 25–30 | 28 | 25 |
| 31–40 | 25 | 30 |
| 41–50 | 26 | 25 |
| 51+ | 4 | 7 |

**Table A.3** *Gender composition of the Aberdare workforce (%)*

|  | Total | | Managers and specialists | | Non-management staff | | Non-management administrators | |
|---|---|---|---|---|---|---|---|---|
|  | 1990 | 1992 | 1990 | 1992 | 1990 | 1992 | 1990 | 1992 |
| Male | 89 | 91 | 100 | 100 | 85 | 88 | 26 | 32 |
| Female | 11 | 9 | – | – | 15 | 12 | 74 | 68 |

of staff with previous experience of cable-making in the company (24 per cent of the workforce in 1990) and many others with high levels of technical skill and experience gained from previous employment. There was no significant variation between the age structures of the three main occupational groups (producers, maintainers, administrators), so their numbers have been aggregated. However, the gender differentiation within the workforce was highly marked, with no female managers or specialists, virtually no female producers or maintainers, but a preponderance of females in non-management administrative grades (see Table A.3).

The final significant characteristic of the workforce was the distance from home to work. If we look at the figures for 1992, over two-thirds of non-management staff lived within 5 km of the factory – that is, in the area in or around Aberdare – and only 7 per cent lived more than 16 km away. In contrast, over two-thirds of managers and specialists lived more than 5 km away from the factory, and 39 per cent lived over 33 km away, usually in more cosmopolitan (e.g. Cardiff) or relatively wealthy rural (e.g. Crickhowell) surroundings. In short, the non-management workforce had a strong vested interest in the survival of the factory, not just in terms of their own jobs but in terms of the prosperity of their local community. The management and specialist workforce, in contrast, had few community ties to the factory, simply job-related ones.

## Observation

Throughout the research, I was given total access to the factory and its staff. Excluding the two five-week interview periods in August/September 1990 and 1992, 27 visits were made to the site between June 1990 and April 1994. Prior to the visits I would always make an appointment with the site personnel officer, but once this had been done I was free to walk around and to stop and speak to staff if they had time and inclination.

The main aim of my more structured observations was to understand the cable-making process and how the various machines and computer systems contributed to the overall production and sale of a cable. Initially, I read various company training manuals and booklets on different cable products and the functions of different parts of the cable-making process (see Chapters 4 and 5). Then, after initial guided tours round the factory by site managers, I asked the· most experienced shift manager if he would recommend to me production staff who were both knowledgeable about the main production areas and able to explain it from a shopfloor point of view to a lay person like myself. After checking with both the relevant shift manager and the staff member, I was then taken by various producers on conducted observational visits to each machine, following through the production of a cable from machine to machine and taking notes as various things were explained to me. Each of these observational visits lasted between 90 minutes and two hours. Subsequently, and over a number of visits, I was introduced by systems engineers to the various elements of the Plant Operations Management System (see Chapters 4, 5 and part of 10), both in the POMS computer room and at an ordinary computer terminal.

## Documentary sources

The company was extremely helpful in giving me access to internal documentation (see the second part of the Bibliography for a list of relevant documents). These documents, plus access to some of the key company files and internal reports, were particularly useful in reconstructing the events of the 1984–8 period. I am also grateful to the managing director of Pirelli Cables Australia for answering numerous questions by letter about the 1984–5 period.

# Bibliography

Company documents are listed in a separate section at the end of the Bibliography

Abernathy, W.J., Clark, K. and Kantrow, A. (1981) 'The new industrial competition', *Harvard Business Review*, Sept./Oct.: 69–81

Abernathy, W.J., Clark, K. and Kantrow, A. (1983) *Industrial Renaissance. Producing a Competitive Future for America*. New York: Basic Books

Adair, J. (1990) *Great Leaders*. Brookwood: Talbot Adair Press

Alexander, L. (1986) 'Successfully implementing strategic decisions', in B. Mayon-White (ed.), *Planning and Managing Change*. London: Harper & Row

Atkinson, J. and Meager, N. (1986) *Changing Working Patterns: How Companies Achieve Flexibility to Meet New Needs*. London: National Economic Development Office

Baldridge, J. (1971) *Power and Conflict in the University*. New York: Wiley

Barnett, C. (1986) *The Audit of War*. London: Macmillan

Bass, B.M. (1990) 'From transactional to transformational leadership: learning to share the vision', *Organizational Dynamics*, 18: 19–31

Bassett, P. (1987) *Strike Free – New Industrial Relations in Britain*. London: Macmillan

Beaumont, P. (1985) 'New plant working practices', *Personnel Review*, 14(1): 15–19

Beaumont, P. and Townley, B. (1985) 'Greenfield sites, new plants and work practices', in V. Hammond (ed.), *Current Research in Management*. London: Frances Pinter

Bedeian, A.G. (1980) *Organization Theory and Analysis*. Hinsdale, IL: Dryden Press

Bessant, J. (1991) *Managing Advanced Manufacturing Technology: the Challenge of the Fifth Wave*. Manchester: NCC/Blackwell

Bessant, J. (1993) 'Towards factory 2000: designing organizations for computer-integrated technologies', in Clark (1993a)

Blau, P.M. (1990) 'Structural constraints and opportunities: Merton's contribution to general theory', in J. Clark, C. Modgil and S. Modgil (eds), *Robert K. Merton – Consensus and Controversy*. London: Falmer Press

Blyton, P. and Turnbull, P. (eds) (1992) *Reassessing Human Resource Management*. London: Sage

Bolland, E. and Goodwin, S. (1988) 'Corporate accounting practice is often a barrier to computer integrated manufacturing', *Industrial Engineering*, 20(7): 24–6

Bonsack, R. (1987) 'Justifying automation in the office and the factory', *Journal of Accounting and EDP*, 3(3): 63–5

*British Business* (1986–9) Selected numbers of weekly journal of the Department of Trade and Industry. London: HMSO

Brown, R. (1991) 'CIM: are we there yet?', *Systems 3X/400*, 19(8): 60–6

Bryman, A. (1986) *Leadership and Organizations*. London: Routledge & Kegan Paul

Bryman, A. (1992) *Charisma and Leadership in Organizations*. London: Sage

Burns, J.M. (1978) *Leadership*. New York: Harper & Row

Campbell-Bradley, I. (1987) *Enlightened Entrepreneurs*. London: Weidenfeld & Nicolson

Carnall, C.A. (1990) *Managing Change in Organizations*. New York: Prentice-Hall

Chandler, A.D. (1962) *Strategy and Structure: Chapters in the History of American Industrial Enterprise*. Cambridge, MA: MIT Press

Child, J. (1972) 'Organizational structure, environment and performance: the role of strategic choice', *Sociology*, 6(1): 2–22

Child, J. (1984) *Organization*. 2nd edn. London: Harper & Row

Cinnamon, A. (1989) 'Why the UK is a tax haven', *Director (UK)*, 43(2): 27–9

Clark, J. (ed.) (1993a) *Human Resource Management and Technical Change*. London: Sage

Clark, J. (1993b) 'Personnel management, human resource management and technical change', in Clark (1993a)

Clark, J. (1993c) 'Full flexibility and self-supervision in an automated factory', in Clark (1993a)

Clark, J. (1993d) 'Managing people in a time of technical change: conclusions and implications', in Clark (1993a)

Clark, J. (1993e) 'Line managers, human resource specialists and technical change', *Employee Relations*, 15(3): 22–8

Clark, J. (1994) 'Computer-integrated manufacturing, supervisory management, and human intervention in the production process', *International Journal of Production Economics*, 34: 305–12

Clark, J., McLoughlin, I., Rose, H. and King, R. (1988) *The Process of Technological Change: New Technology and Social Choice in the Workplace*. Cambridge: Cambridge University Press

Cyert, R.M. and March, J.G. (1963) *A Behavioral Theory of the Firm*. Englewood Cliffs, NJ: Prentice-Hall

Cyert, R.M., Dill, W.R. and March, J.G. (1958) 'The role of expectations in business decision-making', *Administrative Science Quarterly*, 3: 307–40

Dahl, R.A. (1957) 'The concept of power', *Behavioral Science*, 2: 201–15

Daniel, W.W. and Millward, N. (1993) 'Findings from the Workplace Industrial Relations Surveys', in Clark (1993a)

Dawson, P. and Webb, J. (1989) 'New production arrangements: the totally flexible cage?', *Work, Employment and Society*, 3(2): 221–38

Deming, E. (1982) *Quality, Productivity and Competitive Position*. Cambridge, MA: MIT Centre for Advanced Engineering Study

Deming, E. (1986) *Out of the Crisis*. Cambridge, MA: MIT Centre for Advanced Engineering Study

Dent, H. (1990) 'Organizing for the productivity leap: the inevitable automation of management', *Small Business Reports*, 15(9): 31–44

Drucker, P. (1955) *The Practice of Management*. London: Heinemann

Drury, C. (1990) 'Counting the cost of AMT investment (Part I)', *Accountancy*, 105:134–8

Dutton, J. (1988) 'Understanding strategic agenda building and its implications for managing change', in Pondy et al. (1988)

*Economist* (1985) 'Welcome to Wales: a survey', *Economist*, 2 Feb. (supplement)

*Economist* (1991a) 'Britain – regional policy back in fashion', 10 Apr., pp. 56–7

*Economist* (1991b) 'When GM's robots ran amok', 10 Aug.: pp. 64–5

Elger, T. (1991) 'Task flexibility and the intensification of labour in UK manufacturing in the 1980s', in Pollert (1991a)

Fleishman, E.A. (1953a) 'The description of supervisory behavior', *Journal of Applied Psychology*, 37(1): 1–6

Fleishman, E.A. (1953b) 'The measurement of leadership attitudes in industry', *Journal of Applied Psychology*, 37(3): 153–8

Forester, T. (ed.) (1985) *The Information Technology Revolution*. Oxford: Blackwell

Fowler, A. (1987) 'When chief executives discover HRM', *Personnel Management*, 19(3): 3

Garrahan, P. and Stewart, P. (1992) *The Nissan Enigma: Flexibility at Work in a Local Economy*. London: Mansell

Geary, J. (1994) 'Task participation: employees' participation enabled or constrained', in Sisson (1994a)

Gerwin, D. (1984) 'Innovation, microelectronics and manufacturing technology', in M. Warner (ed.), *Microprocessors, Manpower and Society*. Aldershot: Gower

Gottesman, K. (1991) 'JIT manufacturing is more than inventory programs and delivery schedules', *Industrial Engineering*, 23(5): 19–20, 58

Grayson, D. (1986) *The Integrated Payment System in Practice*. London: ACAS, Occasional Paper No. 35

Greenwood, N. (1988) *Implementing Flexible Manufacturing Systems*. London: Macmillan

Grinyer, P., Mayes, D. and McKiernan, P. (1988) *Sharpbenders: The Secrets of Unleashing Corporate Potential*. Oxford: Blackwell

Groover, M. (1987) *Automation, Production Systems, and Computer Integrated Manufacturing*. Englewood Cliffs, NJ: Prentice-Hall

Guest, D. (1989) 'Human resource management: its implications for industrial relations and trade unions', in Storey (1989)

Hague, R. (1989) 'Japanising Geordie-Land?', *Employee Relations*, 11(2): 3–16

Hayes, R.H. and Abernathy, W.J. (1980) 'Managing our way to industrial decline', *Harvard Business Review*, July/Aug.: 69–77

HC (1980) *The Role of the Welsh Office in Developing Employment Opportunities in Wales*. First Report of House of Commons Committee on Welsh Affairs. London: HMSO

HC (1983–4) *The Impact of Regional and Industrial Policy on Wales*. Report of House of Commons Committee on Welsh Affairs. London: HMSO

Hendry, C. (1993) 'Personnel leadership in technical and human resource change', in Clark (1993a)

Hill, S. (1991) 'How do you manage a flexible firm? The total quality model', *Work, Employment and Society*, 5(3): 397–416

Hill, S. and Munday, M. (1991) 'The determinants of inward investment: a Welsh analysis', *Applied Economics*, 23(11): 1761–9

House, R.J. and Mitchell, T.R. (1974) 'Path–goal theory of leadership', *Journal of Contemporary Business*, 3: 81–97

Hrebeniak, L. and Joyce, W. (1984) *Implementing Strategy*. New York: Macmillan

Huang, P. and Sakurai, M. (1990) 'Factory automation: the Japanese experience', *IEEE Transactions on Engineering Management*, 37(2): 102–8

Huczynski, A. and Buchanan, D. (1991) *Organizational Behaviour*. 2nd edn. Hemel Hempstead: Prentice-Hall

Hyman, R. (1991) 'Plus ça change? The theory of production and the production of theory', in Pollert (1991a)

IDS (1992) *Skill-Based Pay*. London: Incomes Data Services, IDS Study No. 500

IEE (1990) *Factory 2001 – Integrating Information and Material Flow*. London: Institution of Electrical Engineers, Conference Publication 323

IEE (1992) *Factory 2000 – Competitive Performance Through Advanced Technology*. London: Institution of Electrical Engineers, Conference Publication 359

IRS (1993a) 'Single-union deals survey: 1', *IRS Employment Trends No. 528*. London: Industrial Relations Services, pp. 3–15

IRS (1993b) 'Single-union deals survey: 2', *IRS Employment Trends No. 529*. London: Industrial Relations Services, pp. 4–12

Ishikawa, K. (1985) *What is Total Quality Control? The Japanese Way*. Englewood Cliffs, NJ: Prentice-Hall

Itami, H. (1987) *Mobilizing Invisible Assets*. Cambridge, MA: Harvard University Press

Jary, S. (1990) 'Trade Union Organization and New Technology Bargaining'. PhD thesis, Department of Sociology and Social Policy, University of Southampton

Johnson, G. and Scholes, K. (1993) *Exploring Corporate Strategy*. 3rd edn. New York: Prentice-Hall

Jones, B. (1988) 'Work and flexible automation in Britain: A review of developments and possibilities', *Work, Employment and Society*, 2(4): 451–86

Juran, J.M. (1989) *Juran on Leadership for Quality*. New York: Free Press

Kanter, R.M. (1983) *The Change Masters*. New York: Simon and Schuster

Katz, H. and Sabel, C. (1985) 'Industrial relations and industrial adjustment in the car industry', *Industrial Relations*, 24(2): 295–315

Kearney (1989) *Computer-Integrated Manufacturing: Competitive Advantage or Technological Dead End?* London: A.T. Kearney Consultants

Keenoy, T. (1990) 'HRM: A case of the wolf in sheep's clothing?', *Personnel Review*, 19(2): 3–9

Kennedy, C. (1991) 'The man who sells Wales to the world', *Director (UK)*, 45(4): 96–111

Kochan, T., Katz, S. and McKersie, R. (1986) *The Transformation of American Industrial Relations*. New York: Basic Books

Kurimoto, A. and So, K. (1990) 'CIM manufacturing strategy', in IEE (1990)

Lawler, E.E. (1978) 'The new plant revolution', *Organizational Dynamics*, Winter: 3–12

Lawler, E.E. (1982) 'Increasing worker involvement to enhance organizational effectiveness', in P.S. Goodman (ed.), *Change in Organizations*. San Francisco: Jossey-Bass

Lawler, E.E. (1986) *High Involvement Management: Participative Strategies for Improving Organizational Performance*. San Francisco: Jossey-Bass

Legge, K. (1978) *Power, Innovation and Problem-Solving in Personnel Management*. London: McGraw-Hill

Legge, K. (1989) 'Human resource management – a critical analysis', in Storey (1989)

Lewis, J. (1988) 'The town that puts industry first', *Business (UK)*, Feb.: 82–4

Lewis, R. (1991) 'Strike-free deals and pendulum arbitration', *British Journal of Industrial Relations*, 28(1): 32–56

Likert, R. (1961) *New Patterns of Management*. New York: McGraw-Hill

Lowe, R. (1990) 'Don't wait for CIM: competitiveness requires more practical approaches', *Industry Week*, 239(12): 77

McCaskey, M.B. (1988) 'The challenge of managing ambiguity and change', in Pondy et al. (1988)

McLoughlin, I. and Clark, J. (1994) *Technological Change at Work*. 2nd edn. Milton Keynes: Open University Press

March, J.G. (1962) 'The business firm as a political coalition', *Journal of Politics*, 24: 662–78

March, J.G. and Simon, H.A. (1958) *Organizations*. New York: Wiley

Marginson, P. and Sisson, K. (1990) 'Single table talk', *Personnel Management*, May: 46–9

Martin, J. (1989) 'Cells drive manufacturing strategy', *Manufacturing Engineering*, 102(1): 49–54

Martin, R. and Jackson, P. (1988) 'Matching advanced manufacturing technology to people', *Personnel Management*, Dec.: 48–51

Meredith, J. and Hill, M. (1987) 'Justifying new manufacturing systems: a managerial approach', *Sloan Management Review*, 28(4): 49–61

Millward, N. (1994) *The New Workplace Industrial Relations*. Aldershot: Dartmouth Publishing

Millward, N., Stevens, M., Smart, D. and Hawes, W.R. (1992) *Workplace Industrial Relations in Transition: The DE/ESRC/PSI/ACAS Surveys*. Aldershot: Dartmouth Publishing

Mintzberg, H. (1973) *The Nature of Managerial Work*. New York: Harper & Row

Mintzberg, H. (1978) 'Patterns in strategy formulation', *Management Science*, 24(9): 934–48

Mintzberg, H. (1990) *Mintzberg on Management*. London: Free Press

Mintzberg, H. and Quinn, J. (eds) (1991) *The Strategy Process*. New York: Prentice-Hall

Mintzberg, H., Raisinghini, M. and Theoret, A. (1976) 'The structure of unstructured decision processes', *Administrative Science Quarterly*, 21: 246–75

*MMC* (1979) *Insulated Electric Wires and Cables*. Monopoly and Mergers' Commission Report. London: HMSO

Moore, J. (1992) *Writers on Strategy and Strategic Management*. Harmondsworth: Penguin

Morita, A. (1994) *Made in Japan*. London: HarperCollins

Morris, J. and Wilkinson, B. (1989) 'Divided Wales'. Report commissioned by HTV Wales for 'Wales This Week'. Cardiff: HTV Wales, mimeo

Morris, J. and Wilkinson, B. (1990) 'Divided Wales: The Cynon Valley Two Years On'. Report commissioned by HTV Wales for 'Wales This Week'. Cardiff: HTV Wales, mimeo

Morris, J. and Wilkinson, B. (1993) 'Poverty and Prosperity in Wales'. Cardiff: Cardiff Business School, mimeo

Newell, H. (1991) 'Field of dreams: evidence of new employee relations on greenfield sites', DPhil, University of Oxford

Newell, H. (1993) 'Exploding the myth of greenfield sites', *Personnel Management*, Jan.: 20–3

Northcott, J. (1986) *Microelectronics in Industry: Promise and Performance*. London: Policy Studies Institute

Oakland, J.S. (1989) *Total Quality Management*. London: Heinemann

Owens, T. (1991) 'The self-managing work team', *Small Business Reports*, 16(2): 53–65

Patterson, W. (1987) 'Computer integrated manufacturing: walking through fire', *Industry Week*, 235(1): 37–44

Pelz, D. and Munson, F. (1982) 'Originality level and the innovating process in organizations', *Human Systems Management*, 3: 173–87

Peters, T. and Waterman, R. (1982) *In Search of Excellence: Lessons from America's Best-Run Companies*. New York: Harper & Row

Pettigrew, A. (1973) *The Politics of Organizational Decision-Making*. London: Tavistock

Pettigrew, A. and Whipp, R. (1991) *Managing Change for Competitive Success*. Oxford: Blackwell

Pettigrew, A., Ferlie, E. and McKee, L. (1992) *Shaping Strategic Change*. London: Sage

Piore, M. and Sabel, C. (1984) *The Second Industrial Divide*. New York: Basic Books

Plossl, K. (1987) *Engineering for the Control of Manufacturing*. Englewood Cliffs, NJ: Prentice-Hall

Pollert, A. (ed.) (1991a) *Farewell to Flexibility?* Oxford: Blackwell

Pollert, A. (1991b) 'The orthodoxy of flexibility', in Pollert (1991a)

Pondy, L., Boland, R. Jr, and Thomas, H. (eds) (1988) *Managing Ambiguity and Change*. New York: Wiley

Popham, P. (1991) 'The best of oriental luck', *Management Today (UK)*, May: 96–7, 104

Porter, M. (1980) *Competitive Strategy*. New York: Free Press

Porter, M. (1981) 'The contributions of industrial organizations to strategic management', *Academy of Management Review*, 6(4): 609–20

Preece, D. (1993) 'Human resource specialists and technical change at greenfield sites', in Clark (1993a)

Purcell, J. and Sisson, K. (1983) 'Strategies and practice in the management of industrial relations', in G.S. Bain (ed.), *Industrial Relations in Britain*. Oxford: Blackwell

Putrus, R. (1991) 'Accounting for intangibles in integrated manufacturing', *Financial and Accounting Systems*, 7(1): 30–5

Quinn, J.B. (1980) *Strategies for Change: Logical Incrementalism*. Homewood, IL: Irwin

Quinn, J.B. (1988) 'Managing strategies incrementally', in J.B. Quinn, H. Mintzberg and R.M. James (eds), *The Strategy Process: Concepts, Contexts and Cases*. Englewood Cliffs, NJ: Prentice-Hall

Robbins, S. (1990) *Organization Theory*. 3rd edn. Englewood Cliffs, NJ: Prentice-Hall

Robbins, T. (1988) *Cults, Converts and Charisma*. London: Sage

Rohan, T. (1987) 'Justifying your CIM investment', *Industry Week*, 235(1): 33–5

RSA (1995) *Tomorrow's Company*. Report of an Inquiry initiated by the Royal Society of Arts. London: RSA

Schein, E. (1983) 'The role of the founder in creating organizational culture', *Organizational Dynamics*, Summer: 13–28

Schein, E. (1985) *Organizational Culture and Leadership*. San Francisco: Jossey-Bass

Sewell, G. and Wilkinson, B. (1992) 'Empowerment or emasculation? Shopfloor surveillance in a total quality organization', in Blyton and Turnbull (1992)

Simon, H.A. (1959) 'Theories of decision-making in economics and behavioral science', *American Economic Review*, 49(3): 253–83

Sims, D., Fineman, S. and Gabriel, Y. (1993) *Organizing and Organizations*. London: Sage

Sisson, K. (1989) 'Personnel management in perspective', in K. Sisson (ed.), *Personnel Management in Britain*. Oxford: Blackwell

Sisson, K. (ed.) (1994a) *Personnel Management*. Oxford: Blackwell

Sisson, K. (1994b) 'Personnel management: paradigms, practice and prospect', in Sisson (1994a)

Small, B. (1992) 'Factory 2000 – translating the vision into reality', in IEE (1992)

Smircich, L. (1983) 'Concepts of culture in organizational analysis', *Administrative Science Quarterly* 28: 339–58

Smith, C., Child, J. and Rowlinson, M. (1990) *Reshaping Work – the Cadbury Experience*. Cambridge: Cambridge University Press

Storey, J. (ed.) (1989) *New Perspectives on Human Resource Management*. London: Routledge

Storey, J. (1992) *Developments in the Management of Human Resources*. Oxford: Blackwell

Storm, D. and Sullivan, S. (1990) 'Justifying strategy for action', *Automation*, 37(8): 40–1

Thomas, H. (1991) 'Grant aid in a bumper package', *Accountancy*, 108: 96–9

Thomsen, S. (1992) 'We are all "Us"', *Columbia Journal of World Business*, 26(4): 6–14

Tichy, N.M. and Devanna, M.A. (1990) *The Transformational Leader*. 2nd edn. New York: Wiley

Trevor, M. (1988) *Toshiba's New British Company – Competitiveness through Innovation in Industry*. London: Policy Studies Institute

Weatherall, D. (1989) 'New technology: IT, JIT and the bottom line', *Management Services* (UK), 33(11): 28–31

Wheatley, M. (1991) 'Milton Keynes: a moving story', *Management Today*, Anniversary Issue: 75–84

Whitaker, A. (1986) 'Managerial strategy and industrial relations: a case study of plant relocation', *Journal of Management Studies*, 23(6): 657–78

Whittington, R. (1993) *What is Strategy – and Does it Matter?* London and New York: Routledge

Whyte, W.F. (1990) 'The new manufacturing organization: problems and opportunities for employee involvement and collective bargaining', *National Productivity Review* (US), 9(3): 337–48

Wickens, P. (1987) *The Road to Nissan: Flexibility, Quality, Teamwork*. London: Macmillan

Wilkinson, A., Marchington, M., Goodman, J. and Ackers, P. (1992) 'Total quality management and employee involvement', *Human Resource Management Journal*, 2(4): 1–20

Wren, C. (1989) 'Factors underlying the employment effects of financial assistance policies', *Applied Economics*, 21(4): 497–513

Wren, C. and Waterson, M. (1991) 'The direct employment effects of financial assistance to industry', *Oxford Economic Papers*, 43(1): 116–38

Yeandle, D. and Clark, J. (1989a) 'A personnel strategy for an automated plant', *Personnel Management*, 21(6): 51–5

Yeandle, D. and Clark, J. (1989b) 'Growing a compatible IR set-up', *Personnel Management*, 21(7): 36–9

## Company Documents and Publications

Bale, C. (1984) 'The Building Wires Project'. Unpublished report to the Managing Director, Pirelli General

*CableTalk* (1992– ) Newsletter of the Energy Cables Division. Eastleigh: Pirelli Cables

Crooks, S.G. (1984) 'Evolution of cable sector organisation and management'. Unpublished paper presented to the Pirelli Cable Sector Chief Executives' Conference, Milan, 21–3 May

*Pirelli Gen* (1982–91) Quarterly Journal of the Pirelli General Group of Companies and Pirelli Construction Company Ltd. Southampton: Pirelli General

Pirelli General (1987) *Employment Handbook – Aberdare Site*. Southampton: Pirelli General

Pirelli General (1988) *Pirelli at Aberdare: Computer Integrated Manufacturing of Cables*. Southampton: Pirelli General

Pirelli General (1989) '75 Years of Pirelli General 1914 to 1989'. Southampton: Pirelli General

Pirelli (1994) 'Welcome to the centre of excellence of automation'. *Fatti e Notizie*. Milan: Pirelli SpA

*TQM Review* (1991– ) Corporate Information Newsletter. Eastleigh: Pirelli Cables

Trezza, G. (1988) 'The plant operations management system', in SEIAF, *Electronic and Information Systems for Factory Automation*. Genoa: Sistemi Elettronici ed Informatici per l'Automazione della Fabbrica

Tynan, O. and James, G. (1985) 'A Report on a Project of Pirelli General UK PLC'. London: ACAS, unpublished report to Pirelli General

# Index

The following abbreviations have been used in the index:
CIM   computer-integrated manufacturing
PG   Pirelli General